Statics and Dynamics
Demystified

Demystified Series

Accounting Demystified
Advanced Statistics Demystified
Algebra Demystified
Alternative Energy Demystified
Anatomy Demystified
asp.net 2.0 Demystified
Astronomy Demystified
Audio Demystified
Biology Demystified
Biotechnology Demystified
Business Calculus Demystified
Business Math Demystified
Business Statistics Demystified
C++ Demystified
Calculus Demystified
Chemistry Demystified
College Algebra Demystified
Corporate Finance Demystified
Databases Demystified
Data Structures Demystified
Differential Equations Demystified
Digital Electronics Demystified
Earth Science Demystified
Electricity Demystified
Electronics Demystified
Environmental Science Demystified
Everyday Math Demystified
Forensics Demystified
Genetics Demystified
Geometry Demystified
Home Networking Demystified
Investing Demystified
Java Demystified
JavaScript Demystified
Linear Algebra Demystified
Macroeconomics Demystified
Management Accounting Demystified

Math Proofs Demystified
Math Word Problems Demystified
Medical Billing and Coding
   Demystified
Medical Terminology Demystified
Meteorology Demystified
Microbiology Demystified
Microeconomics Demystified
Nanotechnology Demystified
Nurse Management Demystified
OOP Demystified
Options Demystified
Organic Chemistry Demystified
Personal Computing Demystified
Pharmacology Demystified
Physics Demystified
Physiology Demystified
Pre-Algebra Demystified
Precalculus Demystified
Probability Demystified
Project Management Demystified
Psychology Demystified
Quality Management Demystified
Quantum Mechanics Demystified
Relativity Demystified
Robotics Demystified
Signals and Systems Demystified
Six Sigma Demystified
sql Demystified
Statics and Dynamics Demystified
Statistics Demystified
Technical Math Demystified
Trigonometry Demystified
uml Demystified
Visual Basic 2005 Demystified
Visual C# 2005 Demystified
xml Demystified

Statics and Dynamics
Demystified

David McMahon

New York   Chicago   San Francisco   Lisbon   London   Madrid
Mexico City   Milan   New Delhi   San Juan   Seoul
Singapore   Sydney   Toronto

Cataloging-in-Publication Data is on file with the Library of Congress.

1 2 3 4 5 6 7 8 9 0 DOC/DOC 0 1 0 9 8 7 6

ISBN 13: 978-0-07-147883-0
ISBN 10: 0-07-147883-3

The sponsoring editor for this book was Judy Bass, the editing supervisor was David E. Fogarty, and the production supervisor was Pamela A. Pelton. It was set in Times Roman by TechBooks. The art director for the cover was Margaret Webster-Shapiro.

Printed and bound by RR Donnelley.

This book is printed on acid-free paper.

McGraw-Hill books are available at special quantity discounts to use as premiums and sales promotions, or for use in corporate training programs. For more information, please write to the Director of Special Sales, McGraw-Hill Professional, Two Penn Plaza, New York, NY 10121-2298. Or contact your local bookstore.

CONTENTS

PREFACE

Statics and Dynamics Demystified is written for anyone who needs help learning the basic concepts and definitions of statics and dynamics. Although geared somewhat for students taking engineering mechanics courses, this book is useful for a broad audience. Students who are taking freshman calculus-based physics, which is mostly about statics and dynamics, will find this book useful although there are some topics here and there that are more advanced than what is covered in that course. Physics students will also find the book useful as a supplement or book to brush up with when taking undergraduate junior level mechanics classes. The last chapter briefly introduces advanced methods like the Lagrangian and Hamiltonian that physics students will find helpful.

The purpose of the book is not to be all encompassing or thorough but instead the idea is to help the reader build a basic foundation for the subject. Therefore, each chapter introduces a few key concepts you will find on a statics and dynamics syllabus. Then example problems are solved. Again, we are not trying to be exhaustively thorough but are instead trying to help students grasp the basic concepts. So, only a few select problems are solved to illustrate the methods used with each type of problem. Chapters are relatively short (compared to what you might find in your textbook) and the end of each chapter has a quiz. For the most part each quiz duplicates the types of problems solved in the text so you can test your knowledge and gain confidence by repeating something you've already seen.

After a review of vector calculus, the book covers basic concepts on forces, gravity, moment of inertia, and friction. This is followed by a study of dynamics which begins with the basic kinematics of particles. Then after a detailed look at Newton's second law, the book examines rotation and circular motion, energy, work and power, and waves and vibrational motion. The book concludes with an introduction to Lagrangian and Hamiltonian methods. It is hoped that engineering students who may not have seen these topics (depending on their particular university) will come away with some new ideas on approaching mechanics.

The book is relatively self-contained, so if you have a modest math background in calculus and differential equations, the book should be useful for self-study.

ABOUT THE AUTHOR

David McMahon is a physicist and researcher at a national laboratory. He is the author of *Linear Algebra Demystified, Quantum Mechanics Demystified, Relativity Demystified,* and *Signals and Systems Demystified.*

CHAPTER 1

A Review of Vector Calculus

Introduction

A vector is a directed line segment between two points P and Q. Vectors have magnitude and direction, and graphically, we represent a vector by an arrow (see Fig. 1-1).

Vectors are represented abstractly by letters. For example, a common vector seen in the study of dynamics is momentum, which is denoted by the letter **p**. In this book, we will denote vectors by boldface type or by placing an arrow on top of the letter used to represent the vector. So **A**, **B**, and $\vec{C}$ are vectors, while a and b represent ordinary numbers which are sometimes called *scalars*. We can perform many familiar algebraic operations on vectors. For instance, we can add two vectors together, to obtain a third vector. Graphically, to add two vectors **A** and **B** to obtain a third vector **C**, we place the tail of **B** at the head of **A**, and then draw a new arrow from the tail of **A** to the head of **B** (see Fig. 1-2).

Fig. 1-1   A vector represented as an arrow.

In order to do calculations, it will be necessary to examine vectors within a particular coordinate system. The *components* of a vector are the *projections* of the vector along the different coordinate axes. In Cartesian coordinates, we use the unit vectors $\hat{i}$, $\hat{j}$, $\hat{k}$ that point along the x, y, and z axes respectively. If we represent the projections of a vector $\vec{A}$ along these axes as $\{a_x, a_y, a_z\}$, then we write $\vec{A}$ as:

$$\vec{A} = a_x\hat{i} + a_y\hat{j} + a_z\hat{k}$$

The numbers $\{a_x, a_y, a_z\}$ are called the *components* of $\vec{A}$ with respect to the basis $\hat{i}$, $\hat{j}$, $\hat{k}$. In a different coordinate system, we will use a different set of basis vectors, and the components of the vector will be different. For now we will stick with the Cartesian coordinate system and adopt the following notation. We write the unit vectors $\hat{i}$, $\hat{j}$, $\hat{k}$ as $\hat{x}$, $\hat{y}$, $\hat{z}$, and thus:

$$\vec{A} = a_x\hat{x} + a_y\hat{y} + a_z\hat{z}$$

Basic Operations on Vectors

The following operations are defined for vectors:

1. Addition
   We add two vectors by adding their components. If:

$$\vec{A} = a_x\hat{x} + a_y\hat{y} + a_z\hat{z} \quad \text{and} \quad \vec{B} = b_x\hat{x} + b_y\hat{y} + b_z\hat{z}$$

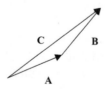

Fig. 1-2   The addition of two vectors.

Then:

$$\vec{C} = \vec{A} + \vec{B} = (a_x + b_x)\hat{x} + (a_y + b_y)\hat{y} + (a_z + b_z)\hat{z}$$

Subtraction is defined in an analogous way.

2. Multiplication by a scalar
   Let a be a real number. Then $a\vec{B}$ is a vector with components defined by:

$$a\vec{B} = a(b_x\hat{x} + b_y\hat{y} + b_z\hat{z}) = ab_x\hat{x} + ab_y\hat{y} + ab_z\hat{z}$$

The vector $a\vec{B}$ is a vector whose length is $|a|$ times the length of $\vec{B}$. If a is positive, then $a\vec{B}$ points in the same direction as $\vec{B}$, while if a is negative, $a\vec{B}$ points in the opposite direction as $\vec{B}$.

3. There exists a zero vector **0** such that:

$$\vec{A} + \mathbf{0} = \vec{A} \text{ for any vector } \vec{A}$$

4. There exists an additive inverse of $\vec{A}$, denoted by $-\vec{A}$, such that:

$$\vec{A} + (-\vec{A}) = \mathbf{0}$$

Multiplication of a vector by another vector is a bit more complicated. In fact, there are two different ways of doing vector multiplication we will consider, the *dot* or *scalar* product that allows us to form the product of two vectors to form a number, and the *cross product* which forms a new vector.

The Dot Product

The dot product between two vectors $\vec{A}$ and $\vec{B}$ is a number and is defined by:

$$\vec{A} \cdot \vec{B} = |A||B|\cos\theta$$

where $|A|$ is the length of $\vec{A}$, $|B|$ is the length of the vector $\vec{B}$, and θ is the angle between the two vectors, taken to be the angle directed from $\vec{A}$ to $\vec{B}$. The dot product can be computed using components in the following way. Again letting $\vec{A} = a_x\hat{x} + a_y\hat{y} + a_z\hat{z}$ and $\vec{B} = b_x\hat{x} + b_y\hat{y} + b_z\hat{z}$, then:

$$\vec{A} \cdot \vec{B} = a_x b_x + a_y b_y + a_z b_z$$

Looking at the angular formula for the dot product, notice that if:

$$\theta = \pi/2, \quad \text{then} \quad \cos\theta = 0$$

$$\theta = 0, \quad \text{then} \quad \cos\theta = 1$$

This tells us that if $\vec{A}$ is perpendicular to $\vec{B}$, and therefore the angle between the two vectors is $\theta = \pi/2$, their dot product is zero.

The dot product can be used to find the length of a vector by taking the dot product of a given vector with itself. In other words:

$$\vec{A} \cdot \vec{A} = |A||A| \cos(0) = |A|^2$$

This tells us how to find the length of a vector from its components:

$$\vec{A} \cdot \vec{A} = a_x^2 + a_y^2 + a_z^2, \quad \Rightarrow \quad |A| = \sqrt{a_x^2 + a_y^2 + a_z^2}$$

The dot product is a linear operation:

$$\vec{A} \cdot (a\vec{B}) = a(\vec{A} \cdot \vec{B})$$

$$\vec{A} \cdot (\vec{B} + \vec{C}) = \vec{A} \cdot \vec{B} + \vec{A} \cdot \vec{C}$$

In addition, the dot product is commutative:

$$\vec{A} \cdot \vec{B} = \vec{B} \cdot \vec{A}$$

EXAMPLE 1-1
Let $\vec{A} = 3\hat{x} - 2\hat{y} + \hat{z}$ and $\vec{B} = 4\hat{x} + 2\hat{y}$.
  (a) Find the length of $\vec{A}$ and the length of $\vec{B}$.
  (b) Compute the dot product between $\vec{A}$ and $\vec{B}$.
  (c) What is the angle between these two vectors?

SOLUTION 1-1
  (a) The length of a vector is found by taking the dot product of the vector with itself. Therefore we find:

$$\vec{A} \cdot \vec{A} = (3)(3) + (-2)(-2) + (1)(1) = 14$$

$$\Rightarrow \quad |A| = \sqrt{14}$$

$$\vec{B} \cdot \vec{B} = (4)(4) + (2)(2) = 20$$

$$\Rightarrow \quad |B| = \sqrt{20}$$

(b) Using the component method, we find:

$$\vec{A} \cdot \vec{B} = (3)(4) + (-2)(2) = 12 - 4 = 8$$

(c) Using the formula:

$$\vec{A} \cdot \vec{B} = |A||B| \cos \theta$$

we find that:

$$\theta = \cos \left(\frac{\vec{A} \cdot \vec{B}}{|A||B|} \right)$$

Using the previous results, we have:

$$\frac{\vec{A} \cdot \vec{B}}{|A||B|} = \frac{8}{\sqrt{14}\sqrt{20}} \approx 0.48$$

And so:

$$\theta = \cos^{-1}(0.48) \cong 61°$$

The Vector Cross Product

The second way to form the product between two vectors is to compute the *cross product*. The result of the cross product is another vector. In Cartesian coordinates, the cross product is computed from the 3×3 determinant:

$$\vec{A} \times \vec{B} = \begin{vmatrix} \hat{x} & \hat{y} & \hat{z} \\ a_x & a_y & a_z \\ b_x & b_y & b_z \end{vmatrix} = \hat{x} \begin{vmatrix} a_y & a_z \\ b_y & b_z \end{vmatrix} - \hat{y} \begin{vmatrix} a_x & a_z \\ b_x & b_z \end{vmatrix} + \hat{z} \begin{vmatrix} a_x & a_y \\ b_x & b_y \end{vmatrix}$$

$$= \hat{x}(a_y b_z - a_z b_y) - \hat{y}(a_x b_z - a_z b_x) + \hat{z}(a_x b_y - a_y b_x)$$

Geometrically, the cross product is given by:

$$\vec{A} \times \vec{B} = |A||B| \sin \theta \, \hat{n}$$

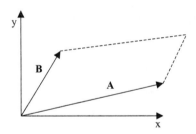

Fig. 1-3 A parallelogram constructed from the cross product of two vectors.

where $\hat{n}$ is a unit vector that points out of the plane formed by $\vec{A}$ and $\vec{B}$. The direction of this unit vector is found from the right hand rule:

1) Point the fingers of your right hand in the direction of $\vec{A}$.

2) Curl your fingers from $\vec{A}$ to $\vec{B}$ along the direction of the smallest angle between the two vectors.

3) Your thumb will be pointing in the direction of $\hat{n}$.

The cross product has the following properties:

1) The cross product is distributive:

$$\vec{A} \times (\vec{B} + \vec{C}) = \vec{A} \times \vec{B} + \vec{A} \times \vec{C}$$

2) It is *not* commutative. In fact:

$$\vec{A} \times \vec{B} = -(\vec{B} \times \vec{A})$$

3) The cross product of any vector $\vec{A}$ with itself is zero:

$$\vec{A} \times \vec{A} = 0$$

Notice that we can construct a parallelogram in the plane from two vectors $\vec{A}$ and $\vec{B}$ (see Fig. 1-3).

The magnitude of the cross product, $|\vec{A} \times \vec{B}|$, is the area of this parallelogram.

Vector Triple Products

The so-called triple products are formed by combining the dot and scalar products in different ways among three vectors. For example, we can form the dot product between a vector $\vec{A}$ and the cross product $\vec{B} \times \vec{C}$. The following identities hold:

$$\vec{A} \cdot (\vec{B} \times \vec{C}) = (\vec{A} \times \vec{B}) \cdot \vec{C}$$
$$\vec{A} \cdot (\vec{B} \times \vec{C}) = \vec{B} \cdot (\vec{C} \times \vec{A})$$

We can also form a triple product by taking the cross product among three vectors. This results in the "bac-cab" rule:

$$\vec{A} \times (\vec{B} \times \vec{C}) = \vec{B}(\vec{A} \cdot \vec{C}) - \vec{C}(\vec{A} \cdot \vec{B})$$

The Cylindrical and Spherical Coordinate Systems

In many problems in electrodynamics, we encounter spherical and cylindrical symmetry. In these cases it is much easier to work in the cylindrical and spherical coordinate systems, respectively.

The cylindrical coordinates are (r,φ,z). They are related to Cartesian coordinates in the following way:

$$x = r \cos \varphi, \quad y = r \sin \varphi, \quad \text{and} \quad z = z$$
$$r = \sqrt{x^2 + y^2}, \quad \varphi = \tan^{-1}(y/x), \quad z = z$$

where $0 \leq \varphi < 2\pi$. Note that in addition to the coordinates themselves, in cylindrical coordinates we work with new basis vectors $(\hat{r}, \hat{\varphi}, \hat{z})$ that are related to the basis vectors in Cartesian coordinates in the following way:

$$\hat{x} = \cos \varphi \, \hat{r} - \sin \varphi \, \hat{\varphi}$$
$$\hat{y} = \sin \varphi \, \hat{r} + \cos \varphi \, \hat{\varphi}$$
$$\hat{z} = \hat{z}$$

The basis vectors in cylindrical coordinates can be written in terms of $\hat{x}, \hat{y}, \hat{z}$ using the following relations:

$$\hat{r} = \cos \varphi \, \hat{x} + \sin \varphi \, \hat{y}$$
$$\hat{\varphi} = -\sin \varphi \, \hat{x} + \cos \varphi \, \hat{y}$$
$$\hat{z} = \hat{z}$$

Spherical coordinates are related to Cartesian coordinates by:

$$x = r \sin\theta \cos\varphi, \quad y = r \sin\theta \sin\varphi, \quad z = r \cos\theta$$

$$r = \sqrt{x^2 + y^2 + z^2}, \quad \theta = \tan^{-1}\left(\frac{\sqrt{x^2 + y^2}}{z}\right), \quad \varphi = \tan^{-1}(y/x)$$

The basis vectors are related in the following way:

$$\hat{x} = \sin\theta \cos\varphi \,\hat{r} + \cos\theta \cos\varphi \,\hat{\theta} - \sin\varphi \,\hat{\varphi}$$

$$\hat{y} = \sin\theta \sin\varphi \,\hat{r} + \cos\theta \sin\varphi \,\hat{\theta} + \cos\varphi \,\hat{\varphi}$$

$$\hat{z} = \cos\theta \,\hat{r} - \sin\theta \,\hat{\theta}$$

$$\hat{r} = \sin\theta \cos\varphi \,\hat{x} + \sin\theta \sin\varphi \,\hat{y} + \cos\theta \,\hat{z}$$

$$\hat{\theta} = \cos\theta \cos\varphi \,\hat{x} + \cos\theta \sin\varphi \,\hat{y} - \sin\theta \,\hat{z}$$

$$\hat{\varphi} = -\sin\varphi \,\hat{x} + \cos\varphi \,\hat{y}$$

The Position Vector

The position vector **r** is a vector that points from the origin (0,0,0) to some point (x,y,z):

$$\vec{r} = x\hat{x} + y\hat{y} + z\hat{z}$$

We can construct a unit vector that points radially outward from the origin by dividing this vector by its magnitude:

$$\hat{r} = \frac{\vec{r}}{|\vec{r}|} = \frac{x\hat{x} + y\hat{y} + z\hat{z}}{\sqrt{x^2 + y^2 + z^2}}$$

In electrodynamics, it will often be necessary to consider two points in a problem. We may need to consider a point **r** where the electric or magnetic field is to be measured, as well as a "source point" **r′** where a source, such as a charge distribution, is located. In such problems we will need to work with a vector that is formed by the difference of these two quantities. Let us denote such a vector by **R**:

$$\vec{R} = \vec{r} - \vec{r}'$$

The unit vector in this direction is given by:

$$\hat{R} = \frac{\vec{r} - \vec{r}'}{|\vec{r} - \vec{r}'|}$$

EXAMPLE 1-2
Let a source point be located at $P_1 = (2, -3, 7)$ and a field point be located at $P_2 = (3, 3, 3)$. Describe the vector that points from the source point to the field point.

SOLUTION 1-2
In this case we have the following vectors:

$$\vec{r} = 3\hat{x} + 3\hat{y} + 3\hat{z}, \quad \vec{r}' = 2\hat{x} - 3\hat{y} + 7\hat{z}$$

Therefore we have:

$$\vec{R} = \vec{r} - \vec{r}' = (3 - 2)\hat{x} + [3 - (-3)]\hat{y} + (3 - 7)\hat{z}$$
$$= \hat{x} + 6\hat{y} - 4\hat{z}$$

The magnitude of this vector is:

$$|\vec{R}| = \sqrt{1^2 + 6^2 + (-4)^2} = \sqrt{1 + 36 + 16} = \sqrt{53}$$

The unit vector that points from the source point to the field point is:

$$\hat{R} = \frac{\vec{R}}{|\vec{R}|} = \frac{\hat{x} + 6\hat{y} - 4\hat{z}}{\sqrt{53}}$$

The Gradient

The *gradient* of a scalar function f maps this function to a vector that we denote by the symbol $\vec{\nabla}f$. The magnitude of this vector at a given point P is the largest directional derivative of f at P. We can make this notion more precise with an example. Let a function $T(x, y, z)$ denote the temperature in a metal plate. The gradient of this function, $\vec{\nabla}T$, tells us in which direction the temperature in the plate increases most rapidly. In Cartesian coordinates, the gradient of a function f is given by:

$$\vec{\nabla}f = \frac{\partial f}{\partial x}\hat{x} + \frac{\partial f}{\partial y}\hat{y} + \frac{\partial f}{\partial z}\hat{z}$$

In many electrodynamics problems it will be necessary to compute the gradient in cylindrical or spherical coordinates. The gradient takes the following form:

$$\vec{\nabla}f = \frac{\partial f}{\partial r}\hat{r} + \frac{1}{r}\frac{\partial f}{\partial \varphi}\hat{\varphi} + \frac{\partial f}{\partial z}\hat{z} \quad \text{(cylindrical coordinates)}$$

$$\vec{\nabla}f = \frac{\partial f}{\partial r}\hat{r} + \frac{1}{r}\frac{\partial f}{\partial \theta}\hat{\theta} + \frac{1}{r\sin\theta}\frac{\partial f}{\partial \varphi}\hat{\varphi} \quad \text{(spherical coordinates)}$$

EXAMPLE 1-3
The temperature in a solid piece of metal is described by the function

$$T(x, y, z) = e^x \sin y \cos z$$

In what direction does the temperature increase most rapidly at the point $P = (0,0,0)$?

SOLUTION 1-3
First, we compute the partial derivatives of T with respect to each of the coordinates, x, y, and z:

$$\frac{\partial T}{\partial x} = \frac{\partial}{\partial x}(e^x \sin y \cos z) = e^x \sin y \cos z$$

$$\frac{\partial T}{\partial y} = \frac{\partial}{\partial y}(e^x \sin y \cos z) = e^x \cos y \cos z$$

$$\frac{\partial T}{\partial z} = \frac{\partial}{\partial z}(e^x \sin y \cos z) = -e^x \sin y \sin z$$

$$\Rightarrow \vec{\nabla}T = e^x \sin y \cos z\,\hat{x} + e^x \cos y \cos z\,\hat{y} - e^x \sin y \sin z\,\hat{z}$$

At $P = (0,0,0)$, we have:

$$\vec{\nabla}T = e^x \sin y \cos z\,\hat{x} + e^x \cos y \cos z\,\hat{y} - e^x \sin y \sin z\,\hat{z}|_{(x,y,z)=(0,0,0)}$$

$$= e^0 \sin(0)\cos(0)\hat{x} + e^0 \cos(0)\cos(0)\hat{y} - e^0 \sin(0)\sin(0)\hat{z}$$

$$= \hat{y}$$

And so, at the point P, the temperature is increasing most rapidly in the y direction.

EXAMPLE 1-4

Compute the gradients of:

(a) $f(x, y, z) = \tan(xy) + \cos x \sin y e^z$

(b) $f(r, \theta, \varphi) = (r^2 \cos \theta + 2r^3) \sin \theta \sin \varphi$ in spherical coordinates

(c) $f(r, \varphi, z) = \dfrac{(5r+1)^4}{7} z + 2 \sin \varphi$ in cylindrical coordinates

SOLUTION 1-4

(a) We compute the partial derivatives of f in the x, y, and z directions:

$$\frac{\partial f}{\partial x} = \frac{\partial}{\partial x}(\tan xy + \cos x \sin y e^z) = y \sec^2 xy - \sin x \sin y e^z$$

$$\frac{\partial f}{\partial y} = \frac{\partial}{\partial y}(\tan xy + \cos x \sin y e^z) = x \sec^2 xy + \cos x \cos y e^z$$

$$\frac{\partial f}{\partial z} = \frac{\partial}{\partial z}(\tan xy + \cos x \sin y e^z) = \cos x \sin y e^z$$

Putting these results together with the form of the gradient in Cartesian coordinates, we obtain:

$$\vec{\nabla} f = \frac{\partial f}{\partial x}\hat{x} + \frac{\partial f}{\partial y}\hat{y} + \frac{\partial f}{\partial z}\hat{z}$$

$$= (y \sec^2 xy - \sin x \sin y e^z)\hat{x} + (x \sec^2 xy + \cos x \cos y e^z)\hat{y}$$

$$+ (\cos x \sin y e^z)\hat{z}$$

(b) In spherical coordinates, the gradient takes the form:

$$\vec{\nabla} f = \frac{\partial f}{\partial r}\hat{r} + \frac{1}{r}\frac{\partial f}{\partial \theta}\hat{\theta} + \frac{1}{r \sin \theta}\frac{\partial f}{\partial \varphi}\hat{\varphi}$$

Computing the partial derivative with respect to r, we obtain:

$$\frac{\partial f}{\partial r} = 2r \cos \theta \sin \theta \sin \varphi + 6r^2 \sin \theta \sin \varphi$$

To compute the derivative with respect to θ, recall the product rule which tells us that:

$$(fg)' = f'g + g'f$$

Therefore:

$$\frac{\partial f}{\partial \theta} = -r^2\sin^2\theta\sin\varphi + r^2\cos^2\theta\sin\varphi + 2r^3\cos\theta\sin\varphi$$

$$= r^2\sin\varphi(\cos^2\theta - \sin^2\theta) + 2r^3\cos\theta\sin\varphi$$

We can simplify this by noting the half-angle identities:

$$\cos^2\theta = \frac{1+\cos 2\theta}{2}, \quad \sin^2\theta = \frac{1-\cos 2\theta}{2}$$

$$\Rightarrow \quad \cos^2\theta - \sin^2\theta = \frac{1}{2} + \frac{\cos 2\theta}{2} - \frac{1}{2} + \frac{\cos 2\theta}{2} = \cos 2\theta$$

And so, we have:

$$\frac{\partial f}{\partial \theta} = r^2\cos 2\theta\sin\varphi + 2r^3\cos\theta\sin\varphi$$

Finally, the last derivative is:

$$\frac{\partial f}{\partial \varphi} = r^2\cos\theta\sin\theta\cos\varphi + 2r^3\sin\theta\cos\varphi$$

Putting these results together, we obtain the gradient:

$$\vec{\nabla}f = \frac{\partial f}{\partial r}\hat{r} + \frac{1}{r}\frac{\partial f}{\partial \theta}\hat{\theta} + \frac{1}{r\sin\theta}\frac{\partial f}{\partial \varphi}\hat{\varphi}$$

$$= 2r\sin\theta\sin\varphi\,(\cos\theta + 3r)\hat{r} + r\sin\varphi\,(\cos 2\theta + 2r\cos\theta)\hat{\theta}$$

$$+ r\cos\varphi\,(\cos\theta + 2r)\,\hat{\varphi}$$

(c) We compute each of the derivatives with respect to the cylindrical coordinates, (r,φ,z):

$$\frac{\partial f}{\partial r} = \frac{\partial}{\partial r}\left(\frac{(5r+1)^4}{7}z + 2\sin\varphi\right)$$

$$= \frac{\partial}{\partial r}\left(\frac{(5r+1)^4}{7}z\right) = \left(\frac{4}{7}\right)(5)(5r+1)^3 z = \frac{20(5r+1)^3}{7}z$$

The formula for the gradient in spherical coordinates divides the φ derivative by r, and so we compute:

$$\frac{1}{r}\frac{\partial f}{\partial \varphi} = \frac{1}{r}\frac{\partial}{\partial \varphi}\left[\frac{(5r+1)^4}{7}z + 2\sin\varphi\right] = \frac{2\cos\varphi}{r}$$

Finally, for the z-derivative we have:

$$\frac{\partial f}{\partial z} = \frac{\partial}{\partial z}\left[\frac{(5r+1)^4}{7}z + 2\sin\varphi\right] = \frac{(5r+1)^4}{7}$$

The gradient is then found to be:

$$\vec{\nabla}f = \frac{\partial f}{\partial r}\hat{r} + \frac{1}{r}\frac{\partial f}{\partial \varphi}\hat{\varphi} + \frac{\partial f}{\partial z}\hat{z}$$

$$= \frac{20(5r+1)^4}{7}z\,\hat{r} + \frac{2\cos\varphi}{r}\hat{\varphi} + \frac{(5r+1)^4}{7}\hat{z}$$

In the next example, we demonstrate the product rule for the gradient, which states that:

$$\vec{\nabla}(fg) = f\,\vec{\nabla}(g) + g\,\vec{\nabla}(f)$$

EXAMPLE 1-5
Verify the product rule for the gradient for the two functions:

$$f(x, y) = 2\cos x \sin y, \quad g(x, y) = x^2 \sin y$$

SOLUTION 1-5
First, we compute the gradient of the product fg:

$$fg = 2x^2 \cos x \sin^2 y,$$

$$\Rightarrow \vec{\nabla}(fg) = \frac{\partial}{\partial x}(2x^2 \cos x \sin^2 y)\hat{x} + \frac{\partial}{\partial y}(2x^2 \cos x \sin^2 y)\hat{y}$$

$$= (4x \cos x \sin^2 y - 2x^2 \sin x \sin^2 y)\hat{x} + (4x^2 \cos x \cos y \sin y)\hat{y}$$

Now, the first term in the product rule for the gradient is:

$$f(\vec{\nabla}g) = 2\cos x \sin y \left(\frac{\partial g}{\partial x}\hat{x} + \frac{\partial g}{\partial y}\hat{y} \right)$$

$$= 2\cos x \sin y \,(2x \sin y \,\hat{x} + x^2 \cos y \,\hat{y})$$

$$= 4x \cos x \sin^2 y \,\hat{x} + 2x^2 \cos x \cos y \sin y \,\hat{y}$$

For the other term, we find:

$$g(\vec{\nabla}f) = g \left(\frac{\partial f}{\partial x}\hat{x} + \frac{\partial f}{\partial y}\hat{y} \right)$$

$$= x^2 \sin y \,(-2\sin x \sin y \,\hat{x} + 2\cos x \cos y \,\hat{y})$$

$$= -2x^2 \sin x \sin^2 y \,\hat{x} + 2x^2 \cos x \cos y \sin y \,\hat{y}$$

Therefore the product rule for gradients gives:

$$f(\vec{\nabla}g) + g(\vec{\nabla}f) = (4x \cos x \sin^2 y - 2x^2 \sin x \sin^2 y)\hat{x}$$

$$+ (2x^2 \cos x \cos y \sin y + 2x^2 \cos x \cos y \sin y)\hat{y}$$

$$= (4x \cos x \sin^2 y - 2x^2 \sin x \sin^2 y)\hat{x}$$

$$+ (4x^2 \cos x \cos y \sin y)\hat{y}$$

The ∇ Operator and Divergence

In Cartesian coordinates, we write the gradient as:

$$\vec{\nabla}f = \frac{\partial f}{\partial x}\hat{x} + \frac{\partial f}{\partial y}\hat{y} + \frac{\partial f}{\partial z}\hat{z}$$

We can think of this as applying the operator ∇ to the function f. This way, we can think of ∇ as an operator that can stand on its own:

$$\vec{\nabla} = \hat{x}\frac{\partial}{\partial x} + \hat{y}\frac{\partial}{\partial y} + \hat{z}\frac{\partial}{\partial z}$$

Sometimes this is referred to as the "del" operator.

We now have an operator can be thought of as a vector. Let's take the dot product of this operator with some vector function:

$$\vec{F} = F_x\hat{x} + F_y\hat{y} + F_z\hat{z}, \quad \Rightarrow$$

$$\vec{\nabla} \cdot \vec{F} = \left(\hat{x}\frac{\partial}{\partial x} + \hat{y}\frac{\partial}{\partial y} + \hat{z}\frac{\partial}{\partial z} \right) \cdot (F_x\hat{x} + F_y\hat{y} + F_z\hat{z})$$

$$= \frac{\partial F_x}{\partial x} + \frac{\partial F_y}{\partial y} + \frac{\partial F_z}{\partial z}$$

We now have a second way to compute a derivative, this time by forming a scalar out of a vector field. The quantity $\vec{\nabla} \cdot \vec{F}$ is called the *divergence* of **F**. If the divergence is non zero in a given region, this tells us that the region contains sources or sinks of the vector field. A source corresponds to a positive divergence, while a sink corresponds to a negative divergence.

In cylindrical coordinates, the divergence is:

$$\vec{\nabla} \cdot \vec{F} = \frac{1}{r}\frac{\partial}{\partial r}(r F_r) + \frac{1}{r}\frac{\partial F_\varphi}{\partial \varphi} + \frac{\partial F_z}{\partial z}$$

where we have taken $\vec{F} = F_r\hat{r} + F_\varphi\hat{\varphi} + F_z\hat{z}$ to be a vector function in cylindrical coordinates. In spherical coordinates, for a function $\vec{F} = F_r\hat{r} + F_\theta\hat{\theta} + F_\varphi\hat{\varphi}$ the divergence is given by:

$$\vec{\nabla} \cdot \vec{F} = \frac{1}{r^2}\frac{\partial}{\partial r}(r^2 F_r) + \frac{1}{r\sin\theta}\frac{\partial}{\partial\theta}(\sin\theta\, F_\theta) + \frac{1}{r\sin\theta}\frac{\partial F_\varphi}{\partial\varphi}$$

EXAMPLE 1-6
Find the divergence of $\vec{F} = x^2 y\,\hat{x} + 2y\,\hat{y} - z^3\hat{z}$.

SOLUTION 1-6
For this vector function we have:

$$F_x = x^2 y, \quad F_y = 2y, \quad F_z = -z^3$$

And so:

$$\frac{\partial F_x}{\partial x} = \frac{\partial}{\partial x}(x^2 y) = 2xy, \quad \frac{\partial F_y}{\partial y}\frac{\partial}{\partial y}(2y) = 2,$$

$$\frac{\partial F_z}{\partial z} = \frac{\partial}{\partial z}(-z^3) = -3z^2$$

Therefore the divergence is:

$$\vec{\nabla} \cdot \vec{F} = \frac{\partial F_x}{\partial x} + \frac{\partial F_y}{\partial y} + \frac{\partial F_z}{\partial z} = 2xy + 2 - 3z^2$$

EXAMPLE 1-7
In spherical coordinates, find the divergence of:

$$\vec{F} = r^2 \sin\theta\, \hat{r} + r \sin\theta\, \hat{\theta} + r \cos^3\varphi\, \hat{\varphi}$$

SOLUTION 1-7
Let's begin by reminding ourselves the form that the divergence takes in spherical coordinates:

$$\nabla \cdot \vec{F} = \frac{1}{r^2}\frac{\partial}{\partial r}(r^2 F_r) + \frac{1}{r\sin\theta}\frac{\partial}{\partial\theta}(\sin\theta\, F_\theta) + \frac{1}{r\sin\theta}\frac{\partial F_\varphi}{\partial\varphi}$$

Now we take the appropriate derivatives found in each term:

$$\frac{\partial}{\partial r}(r^2 F_r) = \frac{\partial}{\partial r}(r^4 \sin\theta) = 4r^3 \sin\theta$$

$$\frac{\partial}{\partial\theta}(\sin\theta\, F_\theta) = \frac{\partial}{\partial\theta}(r \sin^2\theta) = 2r \sin\theta \cos\theta$$

$$\frac{\partial}{\partial\varphi}(F_\varphi) = \frac{\partial}{\partial\varphi}(r \cos^3\varphi) = -3r \cos^2\varphi \sin\varphi$$

$$\Rightarrow \nabla \cdot \vec{F} = \frac{1}{r^2}(4r^3 \sin\theta) + \frac{1}{r\sin\theta}(2r\sin\theta\cos\theta)$$

$$+ \frac{1}{r\sin\theta}(-3r\cos^2\varphi\sin\varphi)$$

$$= 4r\sin\theta + 2\cos\theta - \frac{3\cos^2\varphi\sin\varphi}{\sin\theta}$$

We can also form the dot product in the reverse order between a vector and the del operator, producing a new operator:

$$\vec{G} = G_x\hat{x} + G_y\hat{y} + G_z\hat{z},$$

$$\Rightarrow \vec{G}\cdot\vec{\nabla} = (G_x\hat{x} + G_y\hat{y} + G_z\hat{z})\cdot\left(\hat{x}\frac{\partial}{\partial x} + \hat{y}\frac{\partial}{\partial y} + \hat{z}\frac{\partial}{\partial z}\right)$$

$$= G_x\frac{\partial}{\partial x} + G_y\frac{\partial}{\partial y} + G_z\frac{\partial}{\partial z}$$

If we had $\vec{G} = x^2\hat{x} + 2yz\,\hat{y} - \cos z\,\hat{z}$, $\vec{F} = \sin x \sin y\,\hat{x} - e^x\cos y\,\hat{y} + 2x^3z^4\hat{z}$, then

$$(\vec{G} \cdot \vec{\nabla})\vec{F} = (x^2)\frac{\partial}{\partial x}(\sin x \sin y\,\hat{x} - e^x \cos y\,\hat{y} + 2x^3z^4\hat{z})$$

$$+ 2yz\frac{\partial}{\partial y}(\sin x \sin y\,\hat{x} - e^x \cos y\,\hat{y} + 2x^3z^4\hat{z})$$

$$- \cos z\frac{\partial}{\partial z}(\sin x \sin y\,\hat{x} - e^x \cos y\,\hat{y} + 2x^3z^4\hat{z})$$

$$= x^2(\cos x \sin y\,\hat{x} - e^x \cos y\,\hat{y} + 6x^2z^4\hat{z})$$

$$+ 2yz\,(\sin x \cos y\,\hat{x} + e^x \sin y\,\hat{y}) - \cos z(8x^3z^3\hat{z})$$

$$= (x^2 \cos x \sin y + 2yz \sin x \cos y)\hat{x}$$

$$- (x^2e^x \cos y - 2yze^x \sin y)\hat{y} - (6x^4z^4 + 8x^3z^3 \cos z)\hat{z}$$

The Curl

The curl of a vector field is found by taking the cross product of the del operator with the vector function. In Cartesian coordinates this is:

$$\vec{\nabla} \times \vec{F} = \begin{vmatrix} \hat{x} & \hat{y} & \hat{z} \\ \dfrac{\partial}{\partial x} & \dfrac{\partial}{\partial y} & \dfrac{\partial}{\partial z} \\ F_x & F_y & F_z \end{vmatrix}$$

$$= \hat{x}\left(\frac{\partial F_z}{\partial y} - \frac{\partial F_y}{\partial z}\right) - \hat{y}\left(\frac{\partial F_z}{\partial x} - \frac{\partial F_x}{\partial z}\right) + \hat{z}\left(\frac{\partial F_y}{\partial x} - \frac{\partial F_x}{\partial y}\right)$$

The curl of a vector field is another vector field. Some important identities involving the curl are:

1. The divergence of any curl is zero:

$$\vec{\nabla} \cdot (\vec{\nabla} \times \vec{F}) = 0$$

2. The curl of a gradient is zero:

$$\vec{\nabla} \times \vec{\nabla}f = 0$$

Physically, the curl tells us how much a vector field "swirls" around a given point.

In cylindrical coordinates, the curl is given by:

$$\vec{\nabla} \times \vec{F} = \left(\frac{1}{r} \frac{\partial F_z}{\partial \varphi} - \frac{\partial F_\varphi}{\partial z} \right) \hat{r} + \left(\frac{\partial F_r}{\partial z} - \frac{\partial F_z}{\partial r} \right) \hat{\varphi}$$

$$+ \frac{1}{r} \left(\frac{\partial}{\partial r}(rF_\varphi) - \frac{\partial F_r}{\partial \varphi} \right) \hat{z}$$

and, in spherical coordinates:

$$\vec{\nabla} \times \vec{F} = \frac{1}{r \sin \theta} \left[\frac{\partial}{\partial \theta}(\sin \theta \, F_\varphi) - \frac{\partial F_\theta}{\partial \varphi} \right] \hat{r} + \frac{1}{r} \left[\frac{1}{\sin \theta} \frac{\partial F_r}{\partial \varphi} - \frac{\partial}{\partial r}(rF_\varphi) \right] \hat{\theta}$$

$$+ \frac{1}{r} \left[\frac{\partial}{\partial r}(rF_\theta) - \frac{\partial F_r}{\partial \theta} \right] \hat{\varphi}$$

EXAMPLE 1-8

Find the curl of the vector field

$$\vec{F} = e^x \sin z \, \hat{x} + 2y^2 \, \hat{y} - 3z \cos z \, \hat{z}$$

SOLUTION 1-8

$$\vec{\nabla} \times \vec{F} = \hat{x} \left(\frac{\partial F_z}{\partial y} - \frac{\partial F_y}{\partial z} \right) - \hat{y} \left(\frac{\partial F_z}{\partial x} - \frac{\partial F_x}{\partial z} \right) + \hat{z} \left(\frac{\partial F_y}{\partial x} - \frac{\partial F_x}{\partial y} \right)$$

$$= \hat{x} \left[\frac{\partial}{\partial y}(-3z \cos z) - \frac{\partial}{\partial z}(2y_2) \right] - \hat{y} \left[\frac{\partial}{\partial x}(-3z \cos z) - \frac{\partial}{\partial z}(e^x \sin z) \right]$$

$$+ \hat{z} \left[\frac{\partial}{\partial x}(2y^2) - \frac{\partial}{\partial y}(e^x \sin z) \right] = e^x \cos z \, \hat{y}$$

The Laplacian

The Laplacian is formed by applying the del operator twice. We can act the Laplacian operator on a scalar function or a vector function. The divergence of the gradient of a scalar function gives us a new scalar function, which we compute using second derivatives:

$$\nabla^2 f = \frac{\partial^2 f}{\partial x^2} + \frac{\partial^2 f}{\partial y^2} + \frac{\partial^2 f}{\partial z^2}$$

In cylindrical and spherical coordinates, the Laplacian is given by:

$$\nabla^2 f = \frac{1}{r}\frac{\partial}{\partial r}\left(r\frac{\partial f}{\partial r}\right) + \frac{1}{r^2}\frac{\partial^2 f}{\partial \varphi^2} + \frac{\partial^2 f}{\partial z^2} \quad \text{(cylindrical)}$$

$$\nabla^2 f = \frac{1}{r^2}\frac{\partial}{\partial r}\left(r^2\frac{\partial f}{\partial r}\right) + \frac{1}{r^2 \sin\theta}\frac{\partial}{\partial\theta}\left(\sin\theta \frac{\partial f}{\partial\theta}\right) + \frac{1}{r^2 \sin^2\theta}\frac{\partial^2 f}{\partial\varphi^2} \quad \text{(spherical)}$$

The Laplacian operator can also be applied to a vector field. The result is a new vector field, the Laplacian is applied to each individual component. In Cartesian coordinates:

$$\nabla^2 \vec{F} = \nabla^2 F_x\, \hat{x} + \nabla^2 F_y\, \hat{y} + \nabla^2 F_z\, \hat{z}$$

EXAMPLE 1-9
Find the Laplacian of:
(a) $f(x, y) = 3x^2 y - \sin x \cos y$
(b) $g(r, \varphi, z) = r \sin\varphi - z^3 e^{-2r}$

SOLUTION 1-9
(a) Computing the first and second derivatives of the function, we find:

$$\frac{\partial f}{\partial x} = 6xy - \cos x \cos y, \quad \frac{\partial f}{\partial y} = 3x^2 + \sin x \sin y$$

$$\frac{\partial^2 f}{\partial x^2} = 6y + \sin x \cos y, \quad \frac{\partial^2 f}{\partial y^2} = \sin x \cos y$$

And so the Laplacian is:

$$\nabla^2 f = \frac{\partial^2 f}{\partial x^2} + \frac{\partial^2 f}{\partial y^2} = 6y + \sin x \cos y + \sin x \cos y$$

$$= 6y + 2\sin x \cos y$$

(b) In cylindrical coordinates the Laplacian is:

$$\nabla^2 f = \frac{1}{r}\frac{\partial}{\partial r}\left(r\frac{\partial f}{\partial r}\right) + \frac{1}{r^2}\frac{\partial^2 f}{\partial \varphi^2} + \frac{\partial^2 f}{\partial z^2}$$

Applying this formula to $g(r, \varphi, z) = r\sin\varphi - z^3 e^{-2r}$ we find:

$$\frac{\partial g}{\partial r} = \sin\varphi + 2z^3 e^{-2r}, \quad \frac{\partial g}{\partial \varphi} = r\cos\varphi, \quad \frac{\partial g}{\partial z} = -3z^2 e^{-2r}$$

Therefore, the Laplacian of g is:

$$\nabla^2 g = \frac{1}{r}\frac{\partial}{\partial r}(r\sin\varphi + 2rz^3 e^{-2r}) + \frac{1}{r^2}(-r\sin\varphi) - 6ze^{-2r}$$

$$= \frac{1}{r}(\sin\varphi + 2z^3 e^{-2r} - 4rz^3 e^{-2r}) - \frac{\sin\varphi}{r} - 6ze^{-2r}$$

$$= \frac{2z^3}{r}e^{-2r} - 4z^3 e^{-2r} - 6ze^{-2r}$$

Vector Identities

We now state several vector identities involving the del operator.

$$\vec{\nabla}\times\vec{\nabla}f = 0$$
$$\vec{\nabla}\cdot(\vec{\nabla}\times\vec{F}) = 0$$
$$\vec{\nabla}\times(\vec{\nabla}\times\vec{F}) = \vec{\nabla}(\vec{\nabla}\cdot\vec{F}) - \nabla^2\vec{F}$$
$$\vec{\nabla}(fg) = f(\vec{\nabla}g) + g(\vec{\nabla}f)$$
$$\vec{\nabla}(\vec{F}\cdot\vec{G}) = (\vec{F}\cdot\vec{\nabla})\vec{G} + (\vec{G}\cdot\vec{\nabla})\vec{F} + \vec{F}\times(\vec{\nabla}\times\vec{G}) + \vec{G}\times(\vec{\nabla}\times\vec{F})$$
$$\vec{\nabla}\cdot(g\vec{F}) = \vec{F}\cdot(\vec{\nabla}g) + g(\vec{\nabla}\cdot\vec{F})$$
$$\vec{\nabla}\times(g\vec{F}) = g(\vec{\nabla}\times\vec{F}) - \vec{F}\times(\vec{\nabla}g)$$
$$\vec{\nabla}\times(\vec{F}\times\vec{G}) = (\vec{G}\cdot\vec{\nabla})\vec{F} - (\vec{F}\cdot\vec{\nabla})\vec{G} + \vec{F}(\vec{\nabla}\cdot\vec{G}) - \vec{G}(\vec{\nabla}\cdot\vec{F})$$

Let **r** be a vector from the origin to some point (x,y,z). Then:

$$\vec{\nabla}\cdot\vec{r} = 3, \quad \vec{\nabla}\times\vec{r} = 0$$

EXAMPLE 1-10
Show that $\vec{\nabla} \cdot \vec{r} = 3, \quad \vec{\nabla} \times \vec{r} = 0$

SOLUTION 1-10
$$\vec{r} = x\hat{x} + y\hat{y} + z\hat{z}$$

$$\Rightarrow \vec{\nabla} \cdot \vec{r} = \frac{\partial r_x}{\partial x} + \frac{\partial r_y}{\partial y} + \frac{\partial r_z}{\partial z} = \frac{\partial x}{\partial x} + \frac{\partial y}{\partial y} + \frac{\partial z}{\partial z} = 1 + 1 + 1 = 3$$

For the curl we find:

$$\vec{\nabla} \times \vec{r} = \begin{vmatrix} \hat{x} & \hat{y} & \hat{z} \\ \dfrac{\partial}{\partial x} & \dfrac{\partial}{\partial y} & \dfrac{\partial}{\partial z} \\ x & y & z \end{vmatrix}$$

$$= \hat{x}\left(\frac{\partial z}{\partial y} - \frac{\partial y}{\partial z}\right) - \hat{y}\left(\frac{\partial z}{\partial x} - \frac{\partial x}{\partial z}\right) + \hat{z}\left(\frac{\partial y}{\partial x} - \frac{\partial x}{\partial y}\right) = 0$$

Line Integrals

A *line integral* is an integral of a vector function along a curve that is written as:

$$\int_a^b \vec{F} \cdot d\vec{l}$$

where a and b are given points in space. The differential line element is given by the following expressions in Cartesian, cylindrical, and spherical coordinates (see Table 1-1).

If the point $a = b$, then we have an integral around a closed loop. In that case the line integral is written using the following notation:

$$\oint \vec{F} \cdot d\vec{l}$$

Table 1-1   Line elements for familiar coordinate systems

Coordinate System	Differential Line Element
Cartesian	$d\vec{l} = dx\,\hat{x} + dy\,\hat{y} + dz\,\hat{z}$
Cylindrical	$d\vec{l} = dr\,\hat{r} + r d\varphi\,\hat{\varphi} + dz\,\hat{z}$
Spherical	$d\vec{l} = dr\,\hat{r} + r d\theta\,\hat{\theta} + r\sin\theta\,d\varphi\,\hat{\varphi}$

If the integral depends only on the endpoints of the curve, then it is independent of the path used to move from a to b. When this is true we say that the vector field that we are integrating is *conservative*. When integrating a conservative vector field along a closed curve, the following relation holds:

$$\oint \vec{F} \cdot d\vec{l} = 0 \quad \text{(for conservative fields)}$$

EXAMPLE 1-11
Let two points in the (x,y) plane be:

$$a = (0, 0)$$
$$b = (1, 1)$$

A vector field is given by:

$$\vec{F} = xy^2 \hat{x} + x^2 \hat{y}$$

Find $\int_a^b \vec{F} \cdot d\vec{l}$ along the curves:
   (a) the line $y = x$
   (b) the parabola $y = x^2$

SOLUTION 1-11
In Cartesian coordinates we have:

$$d\vec{l} = dx\,\hat{x} + dy\,\hat{y} + dz\,\hat{z}$$

The basis vectors are orthonormal. This means that:

$$\hat{x} \cdot \hat{x} = \hat{y} \cdot \hat{y} = \hat{z} \cdot \hat{z} = 1$$

and all other dot products between basis vectors are zero. Therefore we have:

$$\vec{F} \cdot d\vec{l} = (xy^2\hat{x} + x^2\hat{y}) \cdot (dx\hat{x} + dy\hat{y} + dz\hat{z}) = xy^2 dx + x^2 dy$$

(a) Along the line $y = x$, we have $dy = dx$ and so:

$$\int_a^b \vec{F} \cdot d\vec{l} = \int_a^b xy^2 dx + x^2 dy = \int_0^1 x^3 dx + x^2 dx \int_0^1 x^3 dx$$

$$+ \int_0^1 x^2 dx = \frac{x^4}{4}\Big|_0^1 + \frac{x^3}{3}\Big|_0^1 = \frac{1}{4} + \frac{1}{3} = \frac{7}{12}$$

(b) Along the parabola $y = x^2$, $dy = 2x\,dx$ and we have:

$$\vec{F} \cdot d\vec{l} = xy^2 dx + x^2 dy = x(x^2)^2 dx + x^2 (2x\,dx)$$

$$= x^5 dx + 2x^3 dx = (x^5 + 2x^3)dx$$

And so we obtain:

$$\int_a^b \vec{F} \cdot d\vec{l} = \int_0^1 (x^5 + 2x^3)dx = \frac{1}{6}x^6 \Big|_0^1 + \frac{1}{2}x^4 \Big|_0^1 = \frac{1}{6} + \frac{1}{2} = \frac{2}{3}$$

EXAMPLE 1-12

Let $\vec{F} = x^2 y^3 \hat{x} - 2xy\,\hat{y}$. Find the line integral of this vector function from the point $a = (1,1,0)$ to $b = (2,3,0)$:

  (a) Using the curve that goes from $(1,1,0)$ to $(2,1,0)$, then to $(2,3,0)$.
  (b) Going directly along the line from $(1,1,0)$ to $(2,3,0)$.

SOLUTION 1-12

  (a) First we calculate:

$$\vec{F} \cdot d\vec{l} = (x^2 y^3 \hat{x} - 2xy\,\hat{y}) \cdot (dx\,\hat{x} + dy\,\hat{y} + dz\,\hat{z})$$

$$= x^2 y^3 dx - 2xy\,dy$$

There are two paths along the curve described in Fig. 1-4.

    In part (a), we will integrate along Path 1 and Path 2. Along Path 1, which goes from $(1,1,0)$ to $(2,1,0)$, $y = 1$ does not change (and so $dy = 0$) and therefore we have:

$$\vec{F} \cdot d\vec{l} = x^2 dx, \quad \Rightarrow \int_a^b \vec{F} \cdot d\vec{l} = \int_1^2 x^2 dx$$

$$= \frac{1}{3}x^3 \Big|_1^2 = \frac{8}{3} - \frac{1}{3} = \frac{7}{3}$$

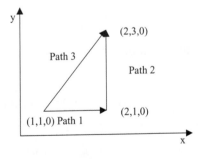

Fig. 1-4   Integration paths for Example 1-12.

On Path 2, $x = 2$ is a constant, therefore $dx = 0$. And so we obtain:

$$\vec{F} \cdot d\vec{l} = -2xy\,dy = -4y\,dy,$$

$$\Rightarrow \int_a^b \vec{F} \cdot d\vec{l} = -\int_1^3 4y\,dy = -2y^2 \Big|_1^3 = -18 + 2 = -16$$

The total answer is found by adding the results obtained on both paths:

$$\int_a^b \vec{F} \cdot d\vec{l} = \frac{7}{3} - 16 = \frac{7}{3} - \frac{48}{3} = -\frac{41}{3}$$

(b) Going directly from $(1,1,0)$ to $(2,3,0)$, which is Path 3 in the diagram, we see that this is a straight line path and therefore is described by some equation of the form $y = mx + b$. At $x = 1$, $y = 2$, and so we find that:

$$2 = m + b$$

While at $x = 2$, $y = 3$, and this gives:

$$3 = 2m + b$$

Subtracting the first equation from the second gives a solution for m:

$$m = 1$$

Substitution of this into the first equation gives:

$$2 = 1 + b$$

therefore, $b = 1$ as well. This allows us to write y in terms of x as:

$$y = x + 1$$

From this equation we deduce that $dy = dx$. Making these substitutions we have:

$$\vec{F} \cdot d\vec{l} = x^2 y^3 dx - 2xy\,dy = x^2(x+1)^3 dx - 2x(x+1)dx$$
$$= x^2(x^3 + 3x^2 + 3x + 1)dx - 2(x^2 + x)dx$$
$$= (x^5 + 3x^4 + 3x^3 - x^2 - 2x)dx$$

We can do the integration in x, and we find that:

$$\int_a^b \vec{F} \cdot d\vec{l} = \int_1^2 (x^5 + 3x^4 + 3x^3 - x^2 - 2x)\, dx$$

$$\times \left. \frac{1}{6}x^6 + \frac{3}{5}x^5 + \frac{3}{4}x^4 - \frac{1}{3}x^3 - x^2 \right|_1^2$$

$$= \frac{64}{6} + \frac{96}{5} + \frac{48}{4} - \frac{8}{3} - 4 - \frac{1}{6} - \frac{3}{5} - \frac{3}{4} + \frac{1}{3} + 1$$

$$= \frac{63}{6} + \frac{93}{5} + \frac{45}{4} - \frac{7}{3} - 3 = \frac{2101}{60}$$

Conservative Vector Fields

A vector field is conservative if:

$$\oint \vec{F} \cdot d\vec{l} = 0$$

for every curve in the region where **F** is defined. If **F** is conservative, then a consequence is that:

$$\vec{\nabla} \times \vec{F} = 0$$

However, the converse is not true in general. If this is true, and since $\vec{\nabla} \times (\vec{\nabla} f) = 0$, this tells us that we can write a conservative vector field in terms of the gradient of some scalar function, i.e.,

$$\vec{F} = \vec{\nabla} f$$

Furthermore, the following is satisfied:

$$\int_a^b \vec{\nabla} f \cdot d\vec{l} = f(b) - f(a)$$

While in general $\vec{\nabla} \times \vec{F} = 0$ does not imply that the vector field **F** is conservative, if the vector field is defined throughout all space, then $\vec{\nabla} \times \vec{F} = 0$ does imply that **F** is conservative.

Statics and Dynamics Demystified

EXAMPLE 1-13

Determine whether or not the vector function:

$$\vec{F} = 2xy\,\hat{x} + x^2\hat{y}$$

is conservative. If so, find a scalar function f such that $\vec{F} = \vec{\nabla}f$.

SOLUTION 1-13

Clearly, $\vec{F}$ is defined everywhere. Therefore we can show $\vec{F}$ is conservative by showing it has zero curl. This vector function has no z-component, and the x and y components do not depend on z, so the curl is simple to evaluate:

$$\vec{\nabla} \times \vec{F} = \left(\frac{\partial F_y}{\partial x} - \frac{\partial F_x}{\partial y} \right)\hat{z} = \left[\frac{\partial}{\partial x}(x^2) - \frac{\partial}{\partial y}(2xy) \right]\hat{z}$$

$$= (2x - 2x)\hat{z} = 0$$

Since the curl of $\vec{F}$ is zero, and it is defined everywhere, it is conservative. Now we compute the anti-derivatives of the x and y components of $\vec{F}$ in order to find the scalar function that satisfies $\vec{F} = \vec{\nabla}f$:

$$\int 2xy\,dx = x^2y + g(y)$$

$$\int x^2\,dy = x^2y + h(x)$$

Now, with no z-dependence, the gradient of f is:

$$\vec{\nabla}f = \frac{\partial f}{\partial x}\hat{x} + \frac{\partial f}{\partial y}\hat{y}$$

Therefore we determine that:

$$f(x, y) = x^2y + C$$

where C is a constant.

EXAMPLE 1-14

Let $T(x, y) = x^2y^3 - 2y$. Construct a vector field from T and show that

$$\int_a^b \vec{\nabla}T \cdot d\vec{l} = T(b) - T(a)$$

For $a = (0,0,0)$ and $b = (3,1,0)$.

SOLUTION 1-14

First, we compute the gradient of T. The partial derivatives of T with respect to x and y are:

$$\frac{\partial T}{\partial x} = \frac{\partial}{\partial x}(x^2y^3 - 2y) = 2xy^3$$

$$\frac{\partial T}{\partial y} = \frac{\partial}{\partial y}(x^2y^3 - 2y) = 3x^2y^2 - 2$$

Now we construct a vector field from T by writing down the gradient:

$$\vec{v} = \vec{\nabla}T = \frac{\partial T}{\partial x}\hat{x} + \frac{\partial T}{\partial y}\hat{y} = 2xy^3\hat{x} + (3x^2y^2 - 2)\hat{y}$$

Now we construct the line integral of this function:

$$\int_a^b \vec{\nabla}T \cdot d\vec{l} = \int_a^b [(2xy^3)\hat{x} + (3x^2y^2 - 2)\hat{y}] \cdot [dx\,\hat{x} + dy\,\hat{y}]$$

$$= \int_a^b (2xy^3)\,dx + (3x^2y^2 - 2)\,dy$$

For the path of integration, we choose the straight line from $(0,0,0)$ to $(3,1,0)$. Therefore we have:

$$Y = (1/3)x, \quad \text{and} \quad dy = (1/3)\,dx$$

We can invert these relationships and integrate along y from 0 to 1:

$$x = 3y, dx = 3\,dy$$

This gives us:

$$\int_a^b 2xy^3\,dx + \int_a^b (3x^2y^2 - 2)\,dy = \int_0^1 2(3y)y^3(3dy) + \int_0^1 (3(3y)^2y^2 - 2)\,dy$$

$$= 18\int_0^1 y^4\,dy + 27\int_0^1 y^4\,dy - 2\int_0^1 dy$$

$$= 45\frac{y^5}{5}\Big|_0^1 - 2y\Big|_0^1 = 9 - 2 = 7$$

Now we wish to verify that:

$$\int_a^b \vec{\nabla} T \cdot d\vec{l} = T(b) - T(a)$$

for $T(x, y) = x^2 y^3 - 2y$.
   At $a = (0,0,0)$, $T = 0$. At $b = (3,1,0)$, we have:

$$T(3,1,0) = (3)^2 (1)^3 - 2(1) = 9 - 2 = 7$$

And so the theorem is verified in this case.

Surface Integrals

We now consider the notion of an integral over a surface. Given a vector field $\vec{F}$, the surface integral over a surface S of $\vec{F}$ is an expression of the form:

$$\int_S \vec{F} \cdot \hat{n} \, da$$

where da is a differential element of area and $\hat{n}$ is a unit vector that is normal to the surface. Over a closed surface, we write this as:

$$\oint_S \vec{F} \cdot \hat{n} \, da$$

Typically, we take this to be the outward normal. In Cartesian coordinates we will consider surfaces in the x-y, y-z, and x-z planes. Table 1-2 shows how to choose the differential element of area and unit normal when performing surface integrals in Cartesian coordinates:

In spherical coordinates, in most cases we will encounter in electrodynamics we will be considering the integration over a sphere at some fixed radius r. In that case r is fixed while the angles φ and θ will vary. In this case we take the

Table 1-2   Differential surface elements in Cartesian coordinates

Integration Surface	Differential Element of Area	Unit Normal
y-z plane (x is fixed)	$dy \, dz$	$\hat{x}$
x-y plane (z is fixed)	$dx \, dy$	$\hat{y}$
x-z plane (y is fixed)	$dx \, dz$	$\hat{z}$

Table 1-3 Differential surface elements in spherical coordinates

Integration Surface	Differential Element of Area	Unit Normal
Surface of sphere (fixed r)	$r^2\sin\theta\,d\theta\,d\varphi$	$\hat{r}$
Fixed θ	$r\,dr\,d\varphi$	$\hat{\theta}$
Fixed φ	$r^2\sin\theta\,dr\,d\theta$	$\hat{\varphi}$

radial vector $\hat{r}$ to be the unit normal. Table 1-3 illustrates the choices used for surface integrals in spherical coordinates.

For surface integration in cylindrical coordinates, we have Table 1-4.

We now consider examples for each of these coordinate systems.

EXAMPLE 1-15
Let $\vec{F} = x^2 z\,\hat{x} + 2yz\,\hat{y} - z^3\,\hat{z}$. Find the surface integral of this function over a unit cube with corner at the origin, i.e., $0 \le x \le 1,\ 0 \le y \le 1, 0 \le z \le 1$.

SOLUTION 1-15
For a cube, there are six surfaces to consider. For the unit cube with a corner at the origin, these are:

1) $x = +1$

2) $x = 0$

3) $y = +1$

4) $y = 0$

5) $z = +1$

6) $z = 0$

To obtain the surface integral over the entire cube, we integrate over each of these six surfaces individually and then add up the result. Starting with surface 1, we have:

$$x = +1,\ dx = 0,\ da = dy\,dz,\ \text{and}\ \hat{n} = \hat{x}$$

And so:

$$\vec{F} \cdot \hat{n} = (x^2 z\,\hat{x} + 2yz\,\hat{y} - z^3\,\hat{z}) \cdot \hat{x} = x^2 z = z\ (\text{for } x = 1)$$

Table 1-4 Differential surface elements in cylindrical coordinates

Integration Surface	Differential Element of Area	Unit Normal
Fixed radius r	$r\,d\varphi\,dz$	$\hat{r}$
Fixed z	$r\,dr\,d\varphi$	$\hat{z}$
Fixed φ	$dr\,dz$	$\hat{\varphi}$

Therefore the integral over this surface is:

$$\int_{S_1} \vec{F} \cdot \hat{n} \, da = \int_0^1 \int_0^1 z \, dy \, dz = \int_0^1 \frac{z^2}{2}\Big|_0^1 dy = \frac{1}{2}\int_0^1 dy = \frac{1}{2}y\Big|_0^1 = \frac{1}{2}$$

Next, at surface 2, we have:

$$\hat{n} = -\hat{x}, \quad \text{however } x = 0, \ \Rightarrow \ \vec{F} \cdot \hat{n} = -x^2 z = 0$$

So the integral over this surface vanishes. Next, we consider surface 3. On this surface,

$$y = +1, \ dy = 0, \ da = dx \, dz, \ \text{and } \hat{n} = \hat{y}$$

And:

$$\vec{F} \cdot \hat{n} = (x^2 z \, \hat{x} + 2yz \, \hat{y} - z^3 \, \hat{z}) \cdot \hat{y} = 2yz = 2z \ \text{(for } y = 1)$$

Setting up the integral, we obtain:

$$\int_{S_3} \vec{F} \cdot \hat{n} \, da = \int_0^1 \int_0^1 2z \, dx \, dz = 2 \int_0^1 \frac{z^2}{2}\Big|_0^1 dx = \int_0^1 dx = x\Big|_0^1 = 1$$

On surface 4, we have:

$$y = 0, \ dy = 0, \ da = dx \, dz, \ \text{and } \hat{n} = -\hat{y}$$

And:

$$\vec{F} \cdot \hat{n} = (x^2 z \, \hat{x} + 2yz \, \hat{y} - z^3 \, \hat{z}) \cdot -\hat{y} = -2yz = 0 \ \text{(for } y = 0)$$

Therefore the integral over this surface is zero. Next we consider surface 5, for which:

$$z = 1, \ dz = 0, \ da = dx \, dy, \ \text{and } \hat{n} = \hat{z}$$

With these conditions we obtain:

$$\vec{F} \cdot \hat{n} = (x^2 z \, \hat{x} + 2yz \, \hat{y} - z^3 \hat{z}) \cdot \hat{z} = -z^3 = -1 \ \text{(for } z = 1)$$

The integral over this surface is found to be:

$$\int_{S_5} \vec{F} \cdot \hat{n} \, da = -\int_0^1 \int_0^1 dx dy = -\int_0^1 x \Big|_0^1 dy$$

$$= -\int_0^1 dy = -y\Big|_0^1 = -1$$

On the last surface, we have $z = 0$. As we found with surfaces 2 and 4, the integral over this surface is zero because the integrand contains z. The final answer is found by adding up the results:

$$\oint_S \vec{F} \cdot \hat{n} \, da = \int_{S_1} \vec{F} \cdot \hat{n} + \int_{S_2} \vec{F} \cdot \hat{n} + \int_{S_3} \vec{F} \cdot \hat{n} + \int_{S_4} \vec{F} \cdot \hat{n} + \int_{S_5} \vec{F} \cdot \hat{n}$$

$$+ \int_{S_6} \vec{F} \cdot \hat{n} = \frac{1}{2} + 1 - 1 = \frac{1}{2}$$

EXAMPLE 1-16
Let $\vec{F} = x\,\hat{x} + y^2\,\hat{y} + z\,\hat{z}$. Compute the surface integral of this function over the surface of a sphere of radius a.

SOLUTION 1-16
First, we rewrite the function in spherical coordinates:

$$\vec{F} = x\,\hat{x} + y^2\,\hat{y} + z\,\hat{z}$$

$$= (r\sin\theta\cos\varphi)(\sin\theta\cos\varphi\,\hat{r} + \cos\theta\cos\varphi\,\hat{\theta} - \sin\varphi\,\hat{\varphi})$$

$$+ (r^2\sin^2\theta\,\sin^2\varphi)(\sin\theta\,\sin\varphi\,\hat{r} + \cos\theta\,\sin\varphi\,\hat{\theta} + \cos\varphi\,\hat{\varphi})$$

$$+ (r\cos\theta)(\cos\theta\,\hat{r} - \sin\theta\,\hat{\theta})$$

$$= (r\sin^2\theta\,\cos^2\varphi + r^2\sin^3\theta\,\sin^3\varphi + r\cos^2\theta)\hat{r}$$

$$+ (r\sin\theta\,\cos\theta\,\cos^2\varphi + r^2\sin^2\theta\,\cos\theta\,\sin^3\varphi - r\cos\theta\,\sin\theta)\hat{\theta}$$

$$- (r\sin\theta\,\cos\varphi\,\sin\varphi - r^2\sin^2\theta\,\sin^2\varphi\,\cos\varphi)\hat{\varphi}$$

$$= r(\sin^2\theta\,\cos^2\varphi + r\sin^3\theta\,\sin^3\varphi + \cos^2\theta)\hat{r}$$

$$+ r\cos\theta\,\sin\theta(\cos^2\varphi + r\sin\theta\,\sin^3\varphi - 1)\hat{\theta}$$

$$+ r\sin\theta\,\cos\varphi\,\sin\varphi(r\sin\theta\,\sin\varphi - 1)\hat{\varphi}$$

Statics and Dynamics Demystified

Referring to the table, for spherical coordinates when performing a surface integral at fixed $r = a$ we take:

$$da = a^2 \sin \theta \, d\theta \, d\varphi, \quad \hat{n} = \hat{r}$$

And so we have:

$$\vec{F} \cdot \hat{n} = F_r = r(\sin^2 \theta \cos^2 \varphi + r \sin^3 \theta \sin^3 \varphi + \cos^2 \theta)$$

The surface integral then takes the following form:

$$\oint \vec{F} \cdot \hat{n} \, da$$

$$= \int_0^{2\pi} \int_0^{\pi} [a(\sin^2 \theta \cos^2 \varphi + r \sin^3 \theta \sin^3 \varphi + \cos^2 \theta)] a^2 \sin \theta d\theta \, d\varphi$$

$$= a^3 \int_0^{2\pi} \int_0^{\pi} \sin^3 \theta \cos^2 \varphi d\theta \, d\varphi + a^4 \int_0^{2\pi} \int_0^{\pi} \sin^4 \theta \sin^3 \varphi \, d\theta \, d\varphi$$

$$+ a^3 \int_0^{2\pi} \int_0^{\pi} \cos^2 \theta \sin \theta \, d\theta d\varphi = (1) + (2) + (3)$$

We consider each term individually, beginning with (1):

$$a^3 \int_0^{2\pi} \int_0^{\pi} \sin^3 \theta \cos^2 \varphi \, d\theta \, d\varphi = a^3 \int_0^{\pi} \sin^3 \theta d\theta \int_0^{2\pi} \cos^2 \varphi d\varphi$$

$$= a^3 \int_0^{\pi} \sin^3 \theta d\theta \int_0^{2\pi} \frac{1 + \cos 2\varphi}{2} \, d\varphi$$

$$= a^3 \int_0^{\pi} \sin^3 \theta \, d\theta \left(\frac{1}{2} \varphi + \frac{1}{4} \sin 2\varphi \right) \Big|_0^{2\pi}$$

$$= a^3 \int_0^{\pi} \sin^3 \theta \, d\theta \, (\pi)$$

Now we use the trig identity $\cos^2 \theta + \sin^2 \theta = 1$ to rewrite the integrand:

$$a^3 \pi \int_0^{\pi} \sin^3 \theta \, d\theta = a^3 \int_0^{\pi} \sin \theta \sin^2 \theta \, d\theta$$

$$= a^3 \pi \int_0^{\pi} \sin \theta (1 - \cos^2 \theta) d\theta$$

$$= a^3 \pi \int_0^{\pi} \sin \theta \, d\theta - a^3 \pi \int_0^{\pi} \sin \theta \cos^2 \theta \, d\theta$$

The first integral can be done immediately:

$$\int_0^\pi \sin\theta d\theta = -\cos\theta\Big|_0^\pi = -\cos\pi + \cos 0 = 1 + 1 = 2$$

For the second integral, we make the substitution $u = \cos\theta$. Then $du = -\sin\theta\, d\theta$ and we have:

$$\int_0^\pi \sin\theta\,\cos^2\theta\,d\theta = -\int_{-1}^1 u^2 du = -\frac{u^3}{3}\Big|_{-1}^1 = -\frac{2}{3}$$

Putting these results together, we find that:

$$a^3\pi \int_0^\pi \sin\theta\,d\theta - a^3\pi \int_0^\pi \sin\theta\cos^2\theta\,d\theta = a^3 2\pi - a^3\frac{2\pi}{3}$$

$$= \frac{4\pi a^3}{3}$$

For the second integral (2) we have:

$$a^4 \int_0^{2\pi}\int_0^\pi \sin^4\theta\,\sin^3\varphi\,d\theta d\varphi = a^4 \int_0^{2\pi}\sin^3\varphi\,d\varphi \int_0^\pi \sin^4\theta d\theta$$

This integral vanishes. Using the previous result:

$$\int_0^{2\pi}\sin^3\varphi\,d\varphi = \int_0^{2\pi}\sin\varphi\,d\varphi - \int_0^{2\pi}\cos^2\varphi\sin\varphi\,d\varphi$$

$$= -\cos\varphi\Big|_0^{2\pi} = -\cos 2\pi + \cos 0 = -1 + 1 = 0$$

To see why $\int_0^{2\pi}\cos^2\varphi\sin\varphi\,d\varphi$ vanishes, note that if we make the substitution $u = \cos\varphi$, the limit at $\varphi = 0$ gives $u = +1$, and the limit $\varphi = 2\pi$ also gives $u = +1$, and so we have:

$$\int_1^1 u^2 du = 0$$

Finally, for (3):

$$(3) = a^3 \int_0^{2\pi} \int_0^\pi \cos^2\theta \sin\theta \, d\theta d\varphi$$

$$= a^3 \int_0^{2\pi} d\varphi \int_0^\pi \cos^2\theta \sin\theta \, d\theta$$

$$= a^3 \left(\varphi \Big|_0^{2\pi} \right) \int_0^\pi \cos^2\theta \sin\theta \, d\theta$$

$$= 2\pi a^3 \int_0^\pi \cos^2\theta \sin\theta \, d\theta$$

Using the same substitution we did previously, we set $u = \cos\theta$, then $du = -\sin\theta d\theta$. At:

$$\theta = 0, u = \cos(0) = +1$$
$$\theta = \pi, u = \cos(\pi) = -1$$

And so the integral becomes:

$$(3) = 2\pi a^3 \int_0^\pi \cos^2\theta \sin\theta \, d\theta$$

$$= -2\pi a^3 \int_1^{-1} u^2 du$$

$$= 2\pi a^3 \int_{-1}^1 u^2 du$$

$$= 2\pi a^3 \frac{1}{3} u^3 \Big|_{-1}^1 = 2\pi a^3 \left(\frac{2}{3} \right) = \frac{4\pi a^3}{3}$$

Adding up $(1) + (2) + (3)$ gives the final result:

$$\oint \vec{F} \cdot \hat{n} \, da = (1) + (2) + (3) = \frac{4\pi a^3}{3} + 0 + \frac{4\pi a^3}{3} = \frac{8\pi a^3}{3}$$

Quiz

1. Let $\vec{A} = \hat{x} + 4\hat{y} + 5\hat{z}$ and $\vec{B} = 2\hat{x} + 2\hat{y}$.
   (a) Find the length of $\vec{A}$ and the length of $\vec{B}$
   (b) Compute the dot product between $\vec{A}$ and $\vec{B}$
   (c) What is the angle between these two vectors?

2. For the vectors $\vec{A} = 4\hat{x} - 2\hat{y} + \hat{z}$ and $\vec{B} = -\hat{x} + 5\hat{y} + 7\hat{z}$ compute $\vec{A} + \vec{B}$, $\vec{A} - \vec{B}$, $|\vec{A}|$ and $|\vec{B}|$.

3. For the vectors $\vec{A} = 4\hat{x} - 2\hat{y} + \hat{z}$ and $\vec{B} = -\hat{x} + 5\hat{y} + 7\hat{z}$ compute $\vec{A} \cdot \vec{B}$. Are these vectors perpendicular?

4. Write the vector $\vec{A} = 4\hat{x} - 2\hat{y} + \hat{z}$ in spherical polar coordinates.

5. The temperature in a solid piece of metal is described by the function

$$T(x, y, z) = xe^{x^2}\sinh y \cos z$$

   In what direction does the temperature increase most rapidly at the point $P = (1,1,1)$?

6. Evaluate

$$\int_0^{2\pi} \left[\int_0^2 \left(\int_3^5 2z^2 r^3 \cos^2 \theta \, dz \right) dr \right] d\theta$$

7. The position of an object of mass m with time is described by $\vec{x} = 4t^2 \hat{i} - 3t \hat{j}$. What is the momentum $\vec{p}$?

8. Determine whether or not the vector function:

$$\vec{F} = 3x^2 y\hat{x} + x^3 \hat{y}$$

   is conservative. If so, find a scalar function f such that $\vec{F} = \vec{\nabla} f$.

9. Let $T(x, y) = 5xy^3 - 2xy$. Construct a vector field from T and show that

$$\int_a^b \vec{\nabla} T \cdot d\vec{l} = T(b) - T(a)$$

   For $a = (0,0,0)$ and $b = (3,1,0)$.

10. Let $\vec{F} = x^2 \hat{x} + 2y \hat{y} + 4z \hat{z}$. Compute the surface integral of this function over the surface of a sphere of radius a.

CHAPTER 2

Particles and Forces

Motion occurs in the universe because forces act on particles causing them to accelerate. We will study motion, in the guise of kinematics and dynamics in more detail later. In this chapter, we will lay out some basic properties of forces.

Characterizing Motion

Before we start, let's think about a few basic ways we can characterize how a body or particle is moving. The first is the position of the body. In one dimension, the position will be a coordinate x that is a function of time

$$x = x(t) \tag{2.1}$$

If the particle is moving in two dimensions, it will be described by the coordinates in the plane $x(t)$ and $y(t)$. Actually, we can describe the motion of the particle along some path by a *position vector* that we denote by $\mathbf{r}(t)$

$$\mathbf{r}(t) = x(t)\mathbf{i} + y(t)\mathbf{j} \tag{2.2}$$

This is a vector that points from the origin to the position of the particle along the path, as illustrated in Fig. 2-1.

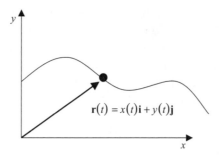

Fig. 2-1   The position vector for the path of a particle.

For the moment, let's stick to one-dimensional motion. The next property we might consider when examining the motion of the particle is how fast is it moving? More precisely, how does the position of the particle vary with time? We compute the derivative of $x(t)$ to answer this question. In one dimension

$$v(t) = \frac{dx}{dt} \tag{2.3}$$

In multiple dimensions, velocity is a vector that not only tells us how fast the particle is going, but also indicates the direction of the velocity. The magnitude of velocity vector is called *speed*.

If the particle or body in question has mass m, then the momentum p is mass times velocity

$$p(t) = mv(t) = m\frac{dx}{dt} \tag{2.4}$$

The rate of change of velocity is the acceleration of the particle

$$a(t) = \frac{dv}{dt} = \frac{d^2x}{dt^2} \tag{2.5}$$

Keep in mind that in general each of these quantities can be a vector. We will review these items again when we study dynamics in more detail.

Newton's First Law

Isaac Newton got physics started in the mid to late 1600s when he discovered his three famous laws. The first of these describes what happens to a particle when no forces act on it.

Common sense tells you that things that aren't being pushed or pulled just sit there. This is basically what the ancients thought—that a body needed a force to be impressed on it in order for it to attain some velocity. But the ancients didn't know much about gravity and frictional forces.

Newton did, and he used this to figure out his first law. Basically the first law says

- If a body is not acted on by a force, it will remain at rest or in a state of uniform motion.

The difference between Newton and the ancients is the last part of that statement—that a body that isn't being acted on by forces and is moving will continue to do so. Think about something floating about in the vacuum of space as an example.

If a body is not being acted on by a force, we can say two more things about it

- The body will have zero *acceleration*.
- If it's in motion, it will move on a straight line.

This tells us that if a body is moving on a curved path, there is a hint that it is under the influence of some force and is therefore accelerating. Conversely, if you know a body is accelerating, then you know it is under the influence of a force. We can find what the force is using Newton's second law.

Newton's Second Law

Newton's second law, which we will investigate in detail in Chapter 9, tells us how to relate force to acceleration. It has a relatively simple form

$$\mathbf{F} = m\mathbf{a} \tag{2.6}$$

We have indicated that force and acceleration are vectors in this equation. Using what we know about acceleration and mass, we can write force in some other ways. Focusing on one-dimensional motion so that we can forget about vectors for a minute, we can also write Newton's second law as the rate of change of momentum

$$F = \frac{dp}{dt} \tag{2.7}$$

Or we can write it in terms of the second derivative of position

$$F = m\frac{d^2x}{dt^2} \qquad (2.8)$$

In this form, given the situation, we can write down differential equations that can be solved to obtain any useful information about the particle we need.

Newton's Third Law

Newton's third law is perhaps the most famous, for it has become part of the common language. This law states that if a body exerts a force on another body, the second body will exert a force on the first that is equal in magnitude but opposite in direction. To every force there is an equal but opposite reaction. It is important to note that the third law does not always apply, so it's not really a law after all.

Couple

Newton's third law leads to a concept in engineering called a *couple*. A couple is two forces that are parallel, with the same magnitude, but with opposite directions.

Deriving Newton's First Law

In fact, the entire content of Newton's laws is really in the second law. Let's say that the force acting on the particle in (2.8) is zero. In this case we solve the differential equation

$$\frac{d^2x}{dt^2} = 0$$

We immediately have that the acceleration is zero. Integrating once gives us an equation for the velocity. We call the constant of integration v_0, giving

$$\frac{dx}{dt} = v_0$$

This tells us that if the particle is at rest, i.e., $v_0 = 0$, it will remain at rest. If $v_0 \neq 0$, this equation tells us that the particle will continue to move at the speed v_0 at all times. What about the path of the particle? We integrate again calling the second constant of integration x_0

$$x(t) = v_0 t + x_0$$

This is the path of the particle—and noticing that it has the form of $y = mx + b$ it's nothing more than a straight line with slope v_0 and intercept x_0.

This little exercise has shown us that Newton's first law is actually contained in the second law for the special case of $F = 0$, in fact we are able to derive the first law from the second mathematically, showing that particles with no forces acting on them move in straight lines and have constant velocity.

We will examine Newton's second law in detail throughout the book, but let's do a couple of examples now to illustrate the process. First, we go one step up from no force to one that is constant.

EXAMPLE 2-1

A particle is moving in one dimension under the influence of a force given by $F(x) = \alpha$ where α is a constant. Find the position and acceleration of the particle. The particle starts from rest.

SOLUTION 2-1

Newton's second law in the two forms (2.6) and (2.8) tells us that we can equate

$$F = m \frac{d^2 x}{dt^2}$$

With the given form of F this equation becomes

$$\frac{d^2 x}{dt^2} = \frac{\alpha}{m}$$

Integrating once we obtain

$$\frac{dx}{dt} = \frac{\alpha}{m} t + v_0 \tag{2.9}$$

We are told that the particle starts from rest. Therefore $v_0 = 0$. We can write down the velocity and momentum of the particle from this expression

$$v(t) = \frac{\alpha}{m} t, \quad p(t) = \alpha t$$

Before finding the position of the particle, let's stop for a moment and think about the constant α. What are the units? Looking at the velocity, we have

$$v(t) = \frac{\alpha}{m}t, \Rightarrow$$

$$[v(t)] = \left[\frac{\alpha}{m}t\right]$$

where we have enclosed each quantity in brackets to indicate we are considering the units or dimensions of each quantity. On the left side

$$[v(t)] = \frac{\text{length}}{\text{time}}$$

On the right side

$$\left[\frac{\alpha}{m}t\right] = [\alpha]\left[\frac{1}{m}\right][t] = [\alpha]\frac{\text{time}}{\text{mass}}$$

We can find the dimensions for the constant by equating these two terms

$$\frac{\text{length}}{\text{time}} = [\alpha]\frac{\text{time}}{\text{mass}}$$

$$\Rightarrow [\alpha] = \left(\frac{\text{mass}}{\text{time}}\right)\left(\frac{\text{length}}{\text{time}}\right) = \frac{\text{mass-length}}{\text{time}^2}$$

In SI units, this is nothing but the force in Newtons

$$[\alpha] = \frac{\text{mass-length}}{\text{time}^2} = \frac{\text{kg-m}}{\text{s}^2} = \text{N}$$

This isn't so surprising of course, since we defined the force to be this constant value. But it's an illustration of how you can check your dimensions and units when in doubt. Continuing, we return to (2.9). Integrating a second time, we find the path of a particle which is under the influence of a constant force α

$$x(t) = \frac{\alpha}{2m}t^2 + v_0 t + x_0$$

We've found that the motion of a particle under the influence of a constant force is a parabola. Since we are told the particle starts from rest, this means that $v_0 = 0$ and in this case $x(t) = \dfrac{\alpha}{2m}t^2 + x_0$.

EXAMPLE 2-2

Suppose now that a particle is moving in one dimension under the influence of a force that varies with time. The force is given by $F(x) = \alpha t$ where α is a constant. Find the position of the particle. The particle starts from rest.

SOLUTION 2-2

Newton's second law in this case becomes

$$\frac{d^2x}{dt^2} = \frac{\alpha}{m}t$$

Integrating, we obtain an expression for the velocity. Since the particle starts from rest we ignore the constant of integration

$$\frac{dx}{dt} = \frac{\alpha}{2m}t^2$$

Integrating a second time gives us the position of the particle

$$x(t) = \frac{\alpha}{6m}t^3 + x_0$$

where x_0 is the initial position of the particle.

EXAMPLE 2-3

A particle is moving with a velocity $v = 3$ m/s. It is not under the influence of any forces. What is the velocity of the particle 2 min later?

SOLUTION 2-3

Newton's first law tells us that the velocity is unchanged, so 2 min later $v = 3$ m/s.

EXAMPLE 2-4

A particle with mass $m = 1$ kg is moving with an initial velocity $v = 3$ m/s. It starts from the origin, under the influence of a constant force $F = 5$ N. What are the position and velocity of the particle after 10 s?

SOLUTION 2-4
The motion of the particle was derived in Example 2-1 where we found that

$$x(t) = \frac{\alpha}{2m}t^2 + v_0 t + x_0$$

With the parameters given in the problem, this becomes

$$x(t) = \frac{5}{2}t^2 + 3t$$

After 10 s, the particle is found at

$$x = \frac{5}{2}(10)^2 + 3(10) = 250 + 30 = 280\,\text{m}$$

In Example 2-2, we found the velocity of the particle is given by the expression

$$v(t) = \frac{\alpha}{m}t + v_0$$

which in this case is

$$v(t) = 5t + 3$$

Hence the speed of the particle at 10 s is

$$v = 5(10) + 3 = 53\,\text{m/s}$$

Line of Action

The *line of action* or *action line* of a force is an imaginary line that corresponds to the force. The line is infinitely long and the vector representing the force can be thought of as a segment of that line.

Forces and Moments

Now that we have introduced some basic ideas, we take a detour to consider *moments*. We have seen that a force tends to distort the path of a particle from a straight line. A moment characterizes the tendency of a force to rotate a body about an axis that passes through the origin. In short a moment is just *torque*. We denote the moment with respect to the origin by $\mathbf{M}$ and define it by

$$\mathbf{M} = \mathbf{r} \times \mathbf{F} \tag{2.10}$$

The position vector from the origin is given by

$$\mathbf{r} = x\,\mathbf{i} + y\,\mathbf{j} + z\,\mathbf{k}$$

A general force in three dimensions is

$$\mathbf{F} = F_x\,\mathbf{i} + F_y\,\mathbf{j} + F_z\,\mathbf{k} \tag{2.11}$$

Then the moment is given by

$$\mathbf{M} = \mathbf{r} \times \mathbf{F} = \begin{vmatrix} \mathbf{i} & \mathbf{j} & \mathbf{k} \\ x & y & z \\ F_x & F_y & F_z \end{vmatrix} \tag{2.12}$$

In this context $\mathbf{r}$ is called the *moment arm*. Since moment is the product of position and force, using SI the units of moment are Newton-meters or N-m. The components of the moment are

$$\begin{aligned} M_x &= yF_z - zF_y \\ M_y &= zF_x - xF_z \\ M_z &= xF_y - yF_x \end{aligned} \tag{2.13}$$

The moment of a couple is found by adding up the moments of each force. Recalling that in a couple, the two forces are the same but oppositely directed. Therefore if $\mathbf{F}_1 = \mathbf{F}$ then

$$\mathbf{F}_2 = -\mathbf{F} \tag{2.14}$$

While the forces in a couple are parallel, the position vectors to each force will not, in general, be parallel. Denoting these vectors by $\mathbf{r}_1$ and $\mathbf{r}_2$, the moment of force $\mathbf{F}_1 = \mathbf{F}$ is

$$\mathbf{M}_1 = \mathbf{r}_1 \times \mathbf{F}_1 = \mathbf{r}_1 \times \mathbf{F}$$

Meanwhile the moment of the second force is

$$\mathbf{M}_2 = \mathbf{r}_2 \times \mathbf{F}_2 = -\mathbf{r}_2 \times \mathbf{F}$$

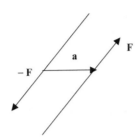

Fig. 2-2   A couple and the distance vector between them.

The moment of the couple is denoted by C and is the sum of the individual moments

$$C = \mathbf{M_1} + \mathbf{M_2} = \mathbf{r_1} \times \mathbf{F} - \mathbf{r_2} \times \mathbf{F} = (\mathbf{r_1} - \mathbf{r_2}) \times \mathbf{F} \qquad (2.15)$$

Mysteriously, the difference between the position vectors $(\mathbf{r_1} - \mathbf{r_2})$ is denoted by $\mathbf{a}$, this might be confusing making you think about acceleration. But keep in mind this is just a position vector. It points from the second position vector to the first, as illustrated in Fig. 2-2.

EXAMPLE 2-5
A force with a magnitude of 100 N is directed from the origin to (3,2,2). Find the scalar moments of this force about the x, y, and z axes.

SOLUTION 2-5
Each component of the force is found by its projection on each of the three axes. The length of the diagonal from the origin to (3,2,2) is

$$d = \sqrt{x^2 + y^2 + z^2} = \sqrt{3^2 + 2^2 + 2^2} = \sqrt{9 + 4 + 4} = \sqrt{17}$$

Then

$$F_x = \frac{x}{d}F = \frac{3}{\sqrt{17}}(100\,\text{N}) = 72.8\,\text{N}$$

And the y and z components of the force are

$$F_y = \frac{y}{d}F = \frac{2}{\sqrt{17}}(100\,\text{N}) = 48.5\,\text{N}$$

$$F_z = \frac{z}{d}F = \frac{2}{\sqrt{17}}(100\,\text{N}) = 48.5\,\text{N}$$

The moment can then be found from the cross product. Since we are asked for the scalar moments of this force about the x, y, and z axes, we can just examine the components (2.13). We have

$$M_x = yF_z - zF_y = (2)(48.5) - (2)(48.5) = 0$$
$$M_y = zF_x - xF_z = (2)(72.8) - (3)(48.5) = 0.1 \text{ N-m}$$
$$M_z = xF_y - yF_x = (3)(48.5) - (2)(72.8) = -0.1 \text{ N-m}$$

The Moment vector is then

$$\mathbf{M} = 0.1\mathbf{j} - 0.1\mathbf{k} \text{ N-m}$$

EXAMPLE 2-6
A force with a magnitude of 75 N is directed from (1,2,1) to (7,3,5). Find the moment of this force acting through the point (2,3,4) with respect to the line passing from (1,−1,−1) to (4,2,3).

SOLUTION 2-6
We can find the components of the moment in two steps. First, we need to find the components of the force. The lengths along each axis are

$$x: 7 - 1 = 6 \text{ m}$$
$$y: 3 - 2 = 1 \text{ m}$$
$$z: 5 - 1 = 4 \text{ m}$$

Now, the force lies along a diagonal which connects the two points. The length of the diagonal is

$$d = \sqrt{(\Delta x)^2 + (\Delta y)^2 + (\Delta z)^2} = \sqrt{6^2 + 1^2 + 4^2} = \sqrt{41}$$

The components of the force are

$$F_x = \frac{\Delta x}{d}F = \frac{6}{\sqrt{41}}(75 \text{ N}) = 70.3 \text{ N}$$

$$F_y = \frac{\Delta y}{d}F = \frac{1}{\sqrt{41}}(75 \text{ N}) = 11.7 \text{ N}$$

$$F_z = \frac{\Delta z}{d}F = \frac{4}{\sqrt{41}}(75 \text{ N}) = 46.9 \text{ N}$$

To find the moment arm we can use a vector that goes from either point on the line to a point which lies on the action line of the force. Let's pick $(1,2,1)$ as the point on the action line of the force. Then using $(1,-1,-1)$, we construct the moment arm by taking the difference of these two points

$$\mathbf{r} = (1-1)\mathbf{i} + (2-(-1))\mathbf{j} + (1-(-1))\mathbf{k} = 3\mathbf{j} + 2\mathbf{k}$$

Then, the components of the moment are

$$M_x = yF_z - zF_y = (3)(46.9) - (2)(11.7) = 117.3 \text{ N-m}$$
$$M_y = zF_x - xF_z = (2)(70.3) - (0)(46.9) = 140.6 \text{ N-m}$$
$$M_z = xF_y - yF_x = (0)(11.7) - (3)(70.3) = -210.9 \text{ N-m}$$

The Moment vector is then

$$\mathbf{M} = 117.3\,\mathbf{i} + 140.6\,\mathbf{j} - 210.9\,\mathbf{k} \text{ N-m}$$

Free Body Diagrams

A *free body diagram* is a way to graphically represent the forces on a body. This can serve as an aid in working out how forces add up. The idea is to draw the force vectors acting in different directions. For example, suppose a body is being acted on by two forces $\mathbf{N}$ and $\mathbf{W}$ as shown in Fig. 2-3. If we also know that the body is not accelerating in the vertical direction, the diagram shows us that

$$\mathbf{N} - \mathbf{W} = 0$$

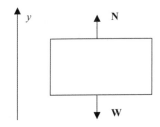

Fig. 2-3   A sample free body diagram.

On the other hand, if the body is accelerating upward, then Newton's second law tells us

$$N - W = ma$$

When solving a problem, start by simply drawing the body and little arrows that represent the forces acting on it. Point the arrows in the direction that the forces are acting.

Quiz

1. A body under the influence of no forces is moving with speed $v = 2$ m/s. What is the speed of the body 2 s later?

2. A body is moving under the influence of a constant force $F = 3$ N. Initially, the body is moving at 1 m/s. What is the speed of the body 5 s later?

3. Write down Newton's second law in terms of acceleration.

4. A particle is moving in one dimension under the influence of a force given by $F \cos \omega t$ where F is a constant. What equation results from Newton's second law?

5. If a particle is acted on by a force given by $F = -kx$ in one dimension, what is the form of the position, $x(t)$?

6. A force with a magnitude of 83 N is directed from the origin to (1,2,3). What are the x, y, and z components of the force?

7. Find the scalar moments of the force in Problem 6 about the x, y, and z axes.

CHAPTER 3

Resultants, Equilibrium, and Statics

In this chapter we go beyond Newton's first law and learn a bit about forces. We begin by applying what we did in Chapter 2 with moments to see how moments can cause a system to move via a resultant force. Then, we examine systems in equilibrium and consider static situations, that is, situations where although forces may be present they balance and cancel each other out so that no motion occurs.

Resultants

The resultant **R** is simply the sum of all the forces that act on a body in the plane. If we denote the n forces acting on a body by F_i where $i = 1, \ldots, n$ then the resultant is

$$\mathbf{R} = \sum_i \mathbf{F}_i \qquad (3.1)$$

Statics and Dynamics Demystified

The magnitude and direction of a resultant are characterized as follows. We calculate the components of the resultant by summing up the x components and the y components of the forces that act on the body, respectively. In other words,

$$R_x = \sum F_x \quad \text{and} \quad R_y = \sum F_y$$

We can then write the resultant vector as

$$\mathbf{R} = \sum F_x \, \mathbf{i} + \sum F_y \, \mathbf{j} \tag{3.2}$$

The magnitude of the resultant is then

$$R = \sqrt{\left(\sum F_x\right)^2 + \left(\sum F_y\right)^2} \tag{3.3}$$

The angle used to characterize the resultant is the angle that the resultant makes with the x axis. This is

$$\tan \theta_x = \frac{\sum F_y}{\sum F_x} \tag{3.4}$$

The *action line* $\bar{a}$ of the resultant is calculated using

$$\bar{a} = \frac{\sum M}{R} \tag{3.5}$$

where $\sum M$ is the sum of the moments acting on the system. If you want a resultant to cause a net moment, you can calculate where the line of action should be, as we will see in the following examples.

EXAMPLE 3-1
Let the following forces act on a body

$\mathbf{F}_1 = 3\mathbf{i} - 2\mathbf{j}$
$\mathbf{F}_2 = 5\mathbf{i} + \mathbf{j}$
$\mathbf{F}_3 = 7\mathbf{i} + 4\mathbf{j}$

Find the resultant and the angle it makes with the x axis. Force is measured in Newtons.

SOLUTION 3-1

We begin by summing the forces

$$\sum F_x = 3 + 5 + 7 = 15 \text{ N}$$

$$\sum F_y = -2 + 1 + 4 = 3 \text{ N}$$

The magnitude of the resultant vector is

$$R = \sqrt{(15)^2 + (3)^2} = 15.3 \text{ N}$$

The angle it makes with the x axis is

$$\theta_x = \tan^{-1}\left(\frac{3}{15}\right) = 11°$$

The resultant vector is

$$\mathbf{R} = 15\,\mathbf{i} + 3\,\mathbf{j}$$

EXAMPLE 3-2

Several forces act on a beam as shown in Fig. 3-1. Find the resultant.

SOLUTION 3-2

The resultant is the sum of the forces. We choose forces acting upward as positive and forces acting down as negative. Then

$$R = \sum F = 200 - 150 + 700 = +750 \text{ N}$$

EXAMPLE 3-3

For the forces in Fig. 3-1, where should the line of force be placed so that the result is an upward moment about O?

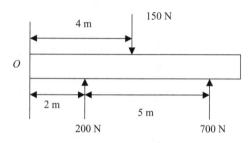

Fig. 3-1   The forces acting on a beam in Examples 3-2 and 3-3.

SOLUTION 3-3

First we sum the moments. The moment arm for each force is the distance in meters of the force from the point O. We find

$$\sum M = 2 \times 200 + 7 \times 700 - 4 \times 150 = 4700 \,\text{N-m}$$

The action line is

$$\bar{a} = \frac{\sum M}{R} = \frac{4700 \,\text{N-m}}{750 \,\text{N}} = 6.3 \,\text{m}$$

Static Equilibrium

Equilibrium is attained by meeting two general conditions. To see what these are, we start with Newton's second law

$$\mathbf{F} = m\mathbf{a}$$

When considering several forces that act on a body, we compute their vector sum and Newton's law becomes

$$\sum \mathbf{F} = m\mathbf{a}$$

If there is an acceleration induced in the body, then motion will occur. Therefore, we see that the first condition for statics is that the vector sum of the forces must vanish

$$\sum \mathbf{F} = 0 \tag{3.6}$$

In order to attain static equilibrium, what we really need to have is that the net vector components of the force must vanish. In other words, the sum of all the components must vanish individually. Therefore (3.6) actually contains three separate conditions within it

$$\sum F_x = 0 \tag{3.7}$$

$$\sum F_y = 0 \tag{3.8}$$

$$\sum F_z = 0 \tag{3.9}$$

In addition to forces, there may be torques (moments) acting on the system. Therefore, a second condition for static equilibrium is that there can be no *net* torques or moments that act on the system. The sum of all the moments must vanish

$$\sum \mathbf{M} = 0 \tag{3.10}$$

Since a moment is a vector, we again require that the sum of the components in all directions must vanish individually. This gives us three more conditions for static equilibrium

$$\sum M_x = 0 \tag{3.11}$$

$$\sum M_y = 0 \tag{3.12}$$

$$\sum M_z = 0 \tag{3.13}$$

When the vector sum of all the forces and all the moments is zero, we say that the system is balanced or is in equilibrium. One simple example is to consider a mass over a pulley. Neglecting friction, there is a downward force due to gravity which is given by $\mathbf{W} = m\mathbf{g}$ where $\mathbf{g}$ is the acceleration due to gravity. If there are no other forces involved, the mass will slide downward.

However, if we exert a tension $\mathbf{T}$ in the chord supporting the mass such that it is pointed in the upward direction and

$$T - mg = 0$$

Then the mass will not move. Equilibrium has been attained in this case.

EXAMPLE 3-4
A boom which is 10 m in length supports a mass $M = 900$ kg. The mass is connected to a chord which is in turn fastened on a wall 5 m away (see Fig. 3-2). Find the tension in the chord and the force on the boom.

SOLUTION 3-4
The system is in equilibrium, therefore the tension in the chord and the force on the boom will be such as to cancel the force of gravity on the mass. First we set up a coordinate system and draw a free body diagram, shown in Fig. 3-3.

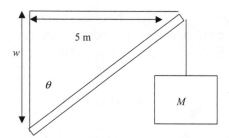

Fig. 3-2 Boom–mass system analyzed in Example 3-4.

The sum of the forces and the torques must sum up to zero in each direction. The force on the mass, caused by gravitation, only acts in the y direction. Therefore the y component of the force on the boom must cancel this force. First, we need to find the angle θ indicated in the figure. As indicated in Fig. 3-4, this is the same angle made between the boom and the wall. We can find the angle if we know the side length of the wall, which is denoted w. We know the length of the boom (10 m—which we take to be the hypotenuse) and the length of the chord, and so the length along the wall is

$$w = \sqrt{(10)^2 - (5)^2} = 8.7 \text{ m}$$

The angle is then defined by

$$\cos \theta = \frac{8.7}{10} = 0.87$$

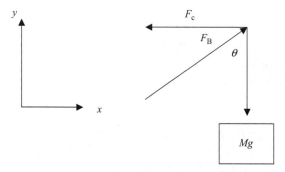

Fig. 3-3 Coordinate system and forces acting in Example 3-4.

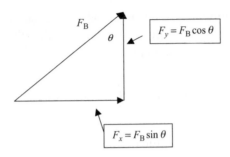

Fig. 3-4   Resolving the force components of the force on the boom.

Now the force due to gravity on the mass is

$$W = -mg = -(900\,\text{kg})(9.81\,\text{m/s}^2) = -8829\,\text{N}$$

We add the minus sign because given our chosen coordinate system, the force points downward. The condition that must be met is

$$\sum F_y = 0$$

Now the y component of the force on the boom, which must point upward to cancel W, can be found by simple trigonometry. This is indicated in Fig. 3-4.

   The side adjacent to the angle is the y component of the boom force. The side opposite the angle is the x component of the force on the boom. The hypotenuse of the triangle is defined by the force on the boom F_B. Since

$$\cos\theta = \frac{\text{adjacent}}{\text{hypotenuse}} \qquad \sin\theta = \frac{\text{opposite}}{\text{hypotenuse}}$$

We have

$$F_x = F_B \sin\theta$$

$$F_y = F_B \cos\theta$$

To meet the condition $\sum F_y = 0$, the following must be satisfied

$$0 = -Mg + F_B \cos\theta$$

$$\Rightarrow F_B = \frac{Mg}{\cos\theta} = \frac{8829}{0.87}\,\text{N} = 10{,}148\,\text{N}$$

The force on the boom has to be larger than the force exerted by the weight of the mass because some of the force on the boom is distributed in the x direction. To find out how much, let's calculate the angle explictly

$$\theta = \cos^{-1}\left(\frac{8.7}{10}\right) = 30°$$

The component of the force on the boom is then

$$F_x = F_B \sin\theta = (10{,}148)\sin 30° = 5074 \text{ N}$$

Now we have everything we need to calculate the force on the chord. We can do it two ways. The easy way is to satisfy

$$\sum F_x = 0$$

Therefore the force on the chord must satisfy

$$F_c + F_x = 0,$$
$$\Rightarrow F_c + 5074 = 0 \quad \text{or} \quad F_c = 5074 \text{ N}$$

To be a bit instructive, let's consider the moments in the problem. The sum of the moments on the system are

$$\sum M = F_c w - (Mg)L$$

where L is the length of the chord. For equilibrium, the system has to satisfy

$$\sum M = 0$$

So we obtain the following equation for the force on the chord

$$0 = \sum M = F_c w - (Mg)L$$
$$= F_c\,(8.7 \text{ m}) - (8829 \text{ N})(5 \text{ m})$$
$$\Rightarrow$$
$$F_c = \frac{(8829 \text{ N})(5 \text{ m})}{8.7 \text{ m}} = 5074 \text{ N}$$

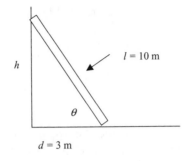

Fig. 3-5 Ladder in Example 3-5.

EXAMPLE 3-5

A ladder is resting with one end on the ground and one end against a wall (see Fig. 3-5). The length of the ladder is $l = 10$ m, while the distance d of the bottom end of the ladder from the wall is 3 m. The ladder has a mass $m_l = 30$ kg. If there is a horizontal frictional force between the ground and the ladder is $F_{gx} = 2$ N but there is no friction between the ladder and the wall, can the ladder support an adult man who crawls up to the midpoint of the ladder?

SOLUTION 3-5

A free body diagram of the forces acting in the problem is shown in Fig. 3-6.

First we use simple trigonometry to determine the height h of the upper part of the ladder. This is

$$h = \sqrt{l^2 - d^2} = \sqrt{(10)^2 - (3)^2} = 9.5 \text{ m}$$

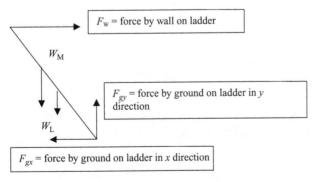

Fig. 3-6 Forces in Example 3-5.

Next, we determine the angle between the ladder and the ground.

$$\sin\theta = \frac{\text{opp}}{\text{hyp}} = \frac{9.5}{10} = 0.95 \Rightarrow$$

$$\theta = \sin^{-1}(0.95) = 72°$$

With the angle in hand, we can find the upward force of the ground on the ladder. We are told that the horizontal frictional force between the ground and the ladder is 2 N. The components of the frictional force satisfy

$$\frac{F_{gy}}{F_{gx}} = \tan\theta$$

Therefore the y component of this force is

$$F_{gy} = F_{gx}\tan\theta = (2\text{ N})\tan 72° = 6.2\text{ N}$$

The sum of the forces in the x direction must balance for equilibrium. Therefore

$$\sum F_x = 0, \Rightarrow$$

$$F_W - F_{gx} = 0$$

From which we conclude the force of the wall on the top of the ladder is 2 N, pointing to the right in Fig. 3-5. To have equilibrium, the sum of the forces in the y direction must also be zero

$$\sum F_y = 0$$

This means that

$$F_{gy} - W_L - W_M = 0$$

The weight of the ladder is $W_L = m_l g = (30\text{ kg})(9.81\text{ m/s}^2) = 294\text{ N}$. From which we find that the weight of the man must satisfy

$$W_M = W_L - F_{gy} = 294 - 6.2 = 288\text{ N}$$

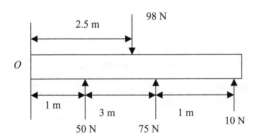

Fig. 3-7　Beam and forces for problems 3–5.

There are about 4.45 N in a pound, so the maximum weight the ladder can support in pounds without slipping is

$$W_M = \frac{288 \text{ N}}{4.45 \text{ N/lb}} = 65 \text{ lb}$$

Under these conditions, the ladder can only support a child, and if an adult man climbed on the ladder the assembly would slip and come tumbling down.

Quiz

1. Given two forces $\mathbf{F_1} = 4\mathbf{i} + 6\mathbf{j}$ and $\mathbf{F_2} = 3\mathbf{i} + 2\mathbf{j}$ find the magnitude of the resultant.

2. For the forces in problem 1, what angle does the resultant make with the *x* axis?

   Several forces act on a beam as shown in Fig. 3-7.

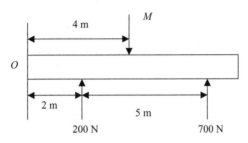

Fig. 3-8　Forces on beam for Problem 3-6.

3. What is the resultant?

4. What is the sum of the moments?

5. Where should the resultant be so as to yield a positive upward moment?

6. Consider the system in Fig. 3-8. By considering both the moments and the forces, what mass M must be placed on the beam to keep the system static? Assume the weight of the beam is 20 N.

CHAPTER 4

Gravity

The force of gravity is one of the most familiar forces. It has been at the heart of science since the birth of physics in the 17th century. Einstein revolutionized the view of gravity with his theory of relativity, but even today gravity remains a subject of intense interest as scientists try to unify gravity with the other known physical forces.

In this chapter we give an overview of Newtonian gravity.

The Gravitational Force

In about the mid-1660s a wave of plague hit England. Newton, who was studying at Cambridge at that time, went to his family farm to escape. While there he was a bit bored and looking for things to do, so he invented calculus, most of mechanics, and the universal law of gravitation, among other things.

To write down the universal law of gravitation we consider two bodies with masses M and m respectively. The force of gravity between the bodies is directed along a line between them and can be written as

$$\mathbf{F} = -G\frac{Mm}{r^2}\hat{\mathbf{r}} \tag{4.1}$$

G is Newton's gravitational constant. In SI units, it has the value

$$G = 6.658 \times 10^{-11}\ \frac{\text{N-m}^2}{\text{kg}^2} \tag{4.2}$$

One of the most important insights that Newton had about the force of gravity, something that was eluding other great minds of the era such as Halley and Hooke, was that the mass of a body looks as though it's all concentrated at its center. This important insight allowed Newton to derive the universal law of gravitation and is known as the *shell theorem*. This theorem states that

- To an outside observer, a uniform shell of mass behaves gravitationally as if all of its mass were concentrated at a point that lies at its center.

Since we are outside the sun, we can treat the sun as if it were a single point with all of its mass concentrated at its center. This is because we can think of the sun or any other body as a series of concentric shells and apply the theorem at any point. So the gist of the shell theorem is when doing gravitational problems, treat the body as a point mass.

EXAMPLE 4-1
In polar coordinates the acceleration vector is given by $\mathbf{a} = (\ddot{r} - r\dot{\phi}^2)\hat{\mathbf{r}} + (r\ddot{\phi} + 2\dot{r}\dot{\phi})\,\hat{\boldsymbol{\phi}}$. Use the universal law of gravitation to say something about the equations of motion of a mass m which is under the influence of the gravitational force of a mass M but experiences no other forces.

SOLUTION 4-1
The gravitational force is given by (4.1). Using Newton's second law $\mathbf{F} = m\mathbf{a}$ we can write

$$m\mathbf{a} = \mathbf{F} = -G\frac{Mm}{r^2}\hat{\mathbf{r}}$$

Using the expression for the acceleration vector in polar coordinates and dividing both sides by the mass m gives

$$(\ddot{r} - r\dot{\phi}^2)\hat{\mathbf{r}} + (r\ddot{\phi} + 2\dot{r}\dot{\phi})\hat{\boldsymbol{\phi}} = -G\frac{M}{r^2}\hat{\mathbf{r}}$$

Vector components must be equal, giving us two equations

$$r\ddot{\phi} + 2\dot{r}\dot{\phi} = 0 \tag{4.3}$$

$$\ddot{r} - r\dot{\phi}^2 = -G\frac{M}{r^2} \tag{4.4}$$

We can say something by examining (4.3) by noticing that

$$\frac{d}{dt}(r^2\dot{\phi}) = 2r\dot{r}\dot{\phi} + r^2\ddot{\phi}$$

And so

$$\frac{1}{r}\frac{d}{dt}(r^2\dot{\phi}) = 2\dot{r}\dot{\phi} + r\ddot{\phi}$$

Therefore (4.3) tells us that

$$\frac{1}{r}\frac{d}{dt}(r^2\dot{\phi}) = 0$$

Multiplying by r

$$\frac{d}{dt}(r^2\dot{\phi}) = 0$$

Which means that $r^2\dot{\phi}$ is a constant, call it C. It can be shown that this constant is proportional to the differential area swept out by the radial vector which points to the particle as it orbits around the mass M

$$\frac{1}{2}r^2\dot{\phi} = \frac{dA}{dt}$$

That is the change in area with time $\frac{dA}{dt}$ is a constant. This is Kepler's law that the radial vector of a body in orbit sweeps out equal areas in equal times.

The path that the particle actually follows in its orbit can be found by solving (4.4). This is easier to do if we make a change of variables

$$r = \frac{1}{u}, \Rightarrow \frac{dr}{du} = -\frac{1}{u^2}$$

The variation of r with time can then be written as

$$\frac{dr}{dt} = \frac{dr}{dt}\frac{du}{du} = \frac{dr}{du}\frac{du}{dt} = -\frac{1}{u^2}\frac{du}{dt}$$

Going a step further

$$-\frac{1}{u^2}\frac{du}{dt} = -\frac{1}{u^2}\frac{du}{dt}\frac{d\phi}{d\phi} = -\frac{1}{u^2}\frac{du}{d\phi}\frac{d\phi}{dt}$$

Returning to what we found earlier, that $r^2\dot{\phi}$ is a constant we call C, we have

$$r^2\dot{\phi} = \frac{1}{u^2}\dot{\phi} = C, \Rightarrow \dot{\phi} = Cu^2$$

So we have arrived at the result

$$\dot{r} = -\frac{1}{u^2}\frac{du}{d\phi}\frac{d\phi}{dt} = -\frac{1}{u^2}\frac{du}{d\phi}Cu^2 = -C\frac{du}{d\phi}$$

Also

$$\ddot{r} = -C^2u^2\frac{d^2u}{d\phi^2}$$

Putting everything together gives a differential equation

$$-C^2u^2\frac{d^2u}{d\phi^2} - C^2u^3 = -GMu^2$$

Which simplifies to

$$\frac{d^2u}{d\phi^2} + u = GM/C^2$$

We try a solution $u = A\cos\phi + B$ where A and B are constants to be determined. Then

$$\frac{du}{d\phi} = -A\sin\phi \text{ and } \frac{d^2u}{d\phi^2} = -A\cos\phi$$

Substitution into our differential equation gives

$$-A\cos\phi + A\cos\phi + B = GM/C^2$$

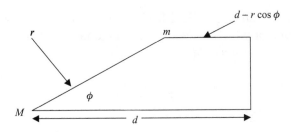

Fig. 4-1 Parameters in a gravitational orbit.

From which we immediately conclude that $B = GM/C^2$. The motion of the mass is described by

$$\frac{1}{r} = u = A \cos \phi + GM/C^2$$

In Fig. 4-1, we lay out some parameters in the orbit.
The eccentricity is defined to be

$$e = \frac{r}{d - r \cos \phi} \tag{4.5}$$

If $e = 0$, the orbit is circular. Real planetary orbits are nearly, but not quite circular. Planets orbit the sun on elliptical paths, which are defined by $0 < e < 1$. If $e > 1$, the orbit is a hyperbola. This is an orbit where the body comes in and approaches the central mass, then skips out never to return again. The values e and d satisfy

$$ed = \frac{C^2}{GM}$$

Giving a solution of the constant C

$$C = \sqrt{GMed} \tag{4.6}$$

Gravitational Acceleration

Near the surface of the earth, the acceleration due to the force of gravity is denoted by g which is a constant with the value

$$g = 9.81 \, \text{m/s}^2 \tag{4.7}$$

in SI units. In U.S. units $g = 32.2 \text{ ft/s}^2$. More generally, the acceleration due to gravity is a vector which is dependent on the radius of the particle from the center of the attracting mass M

$$\mathbf{g} = \frac{GM}{r^2}\hat{\mathbf{r}} \qquad (4.8)$$

The *weight* of a body is the mass m of the body times the acceleration due to gravity. In magnitude

$$W = mg = G\frac{Mm}{r^2} \qquad (4.9)$$

In other words, weight is just the gravitational force that acts on the body. Therefore it is a vector, but in most cases we can just worry about the magnitude since it is purely radial. We can quickly derive (4.7) knowing the radius and mass of the earth. The radius of the earth is $r = 6.37 \times 10^6$ m and the mass of the earth is $M = 5.98 \times 10^{24}$ kg, and so

$$g = \left(6.658 \times 10^{-11}\ \frac{\text{N-m}^2}{\text{kg}^2}\right)\frac{5.98 \times 10^{24}\ \text{kg}}{(6.37 \times 10^6\ \text{m})^2} = 9.81\ \text{m/s}^2$$

EXAMPLE 4-2
What is the acceleration due to gravity on the moon? The mass of the moon is $M = 7.36 \times 10^{22}$ kg and the radius of the moon is $r = 1.74 \times 10^6$ m.

SOLUTION 4-2
Following the example used to find the acceleration due to gravity near the surface of the earth

$$g_{\text{moon}} = \left(6.658 \times 10^{-11}\ \frac{\text{N-m}^2}{\text{kg}^2}\right)\frac{7.36 \times 10^{22}\ \text{kg}}{(1.74 \times 10^6\ \text{m})^2} = 1.62\ \text{m/s}^2$$

EXAMPLE 4-3
A body weighs 170 lb on earth. What is the weight of the body on the moon? Note that 1 lb = 4.45 N.

SOLUTION 4-3
Converting the weight of the body to Newtons

$$(170\ \text{lb})(4.45\ \text{N/lb}) = 757\ \text{N}$$

The mass of the body is

$$m = \frac{W}{g} = \frac{757}{9.81} = 77 \text{ kg}$$

On the moon, this body weighs

$$W_{\text{moon}} = mg_{\text{moon}} = (77 \text{ kg})(9.81 \text{ m/s}^2) = 125 \text{ N}$$

This is about 28 lb, a level of weight loss Jenny Craig can only dream of.

EXAMPLE 4-4
What is the force of gravity between a rock weighing 20 kg and another rock weighing 11 kg that are 1 m apart?

SOLUTION 4-4
The force of gravity is

$$F = \frac{GMm}{r^2}$$

$$= \frac{\left(6.67 \times 10^{-11} \text{ N-m}^2/\text{kg}^2\right)}{(1 \text{ m})^2}$$

$$= 1.47 \times 10^{-8} \text{ N}$$

This is a very small force indeed.

The Law of Periods

The law of periods relates the period of an orbit to the height or radius of the orbit

$$T^2 = \frac{4\pi^2}{GM}r^3 \tag{4.10}$$

If a satellite is orbiting earth at height h and we denote the radius of the earth by R, then the law of periods is

$$T^2 = \frac{4\pi^2}{GM}(R+h)^3$$

where T is the period of the satellite's orbit in seconds if SI units are being used.

EXAMPLE 4-5
A satellite takes 94 min to orbit the earth. What height is it above the earth's surface?

SOLUTION 4-5
The period of the orbit in seconds is

$$(94 \text{ min}) \left(\frac{60 \text{ s}}{\text{min}} \right) = 5640 \text{ s}$$

Rearranging the equation for the period, we have

$$h = \left(\frac{GM}{4\pi^2} T^2 \right)^{1/3} - R$$

$$= \left(\frac{(6.67 \times 10^{-11} \text{ N-m}^2/\text{kg}^2)(5.98 \times 10^{24} \text{ kg})}{4\pi^2}(5640 \text{ s})^2 \right)^{1/3} - (6.37 \times 10^6 \text{ m})$$

$$= 480 \text{ km}$$

The Gravitational Potential

The acceleration due to gravity which is in general a vector

$$\mathbf{g} = -G\frac{M}{r^2}\hat{\mathbf{r}} \tag{4.11}$$

This vector is called the *gravitational field vector*. It can be written in terms of the gradient of a scalar potential. We call this potential the *gravitational potential* ϕ, and the relationship between it and the acceleration due to gravity is

$$\mathbf{g} = -\nabla\phi \tag{4.12}$$

The gravitational field vector is entirely radial. Therefore the gradient given in (4.12) must be entirely radial also. Since

$$\nabla\phi = \frac{d\phi}{dr}\hat{\mathbf{r}} \tag{4.13}$$

in this case, as a result of Newton's law of gravitation, we arrive at a simple differential equation. This equation is

$$\frac{d\phi}{dr} = G\frac{M}{r^2} \tag{4.14}$$

We can solve this equation to obtain an explicit form for the potential of a point mass. Cross multiplication gives

$$d\phi = G\frac{M}{r^2}dr$$

Integrating, we obtain the functional form of the potential which is

$$\phi(r) = -\frac{GM}{r} \tag{4.15}$$

The dimensions of the gravitational potential are

$$[\phi] = \frac{\text{force-length}}{\text{mass}}$$

In SI units, we have

$$[\phi] = \frac{\text{N-m}}{\text{kg}}$$

The potential energy of a mass m in a gravitational field with potential ϕ is

$$V = m\phi \tag{4.16}$$

EXAMPLE 4-6
Find the escape velocity at the surface of the earth.

SOLUTION 4-6
Let m be the mass of the particle. If the particle is moving at speed v, then the kinetic energy which we denote as T is given by

$$T = \frac{1}{2}mv^2$$

The potential energy can be written using (4.15)

$$V = -G\frac{Mm}{R}$$

where M is the mass of the earth and R is the radius of the earth. Conservation of energy dictates that

$$T + V = 0$$

Hence

$$\frac{1}{2}mv^2 = \frac{GMm}{r}$$

Notice the mass of the particle cancels. The escape velocity is then found to be

$$v = \sqrt{\frac{2GM}{R}} = \sqrt{\frac{2\,(6.67 \times 10^{-11})\,(5.98 \times 10^{24})}{6.37 \times 10^6}} = 11.2 \text{ km/s}$$

If the mass density of a body is ρ then the gravitational potential is found by adding up or integrating over the entire body. That is, it is given by the volume integral

$$\phi = -G \int_V \frac{\rho}{r}\, dV \qquad (4.17)$$

Finally, Poisson's equation is a differential equation that can be used to relate the potential to the mass density at a point.

$$\nabla^2 \phi = 4\pi\, G\rho \qquad (4.18)$$

EXAMPLE 4-7
Suppose a mass m is located on axis a height z above a thin disk with radius R. A total mass M is contained on the disk and is distributed uniformly. Find the gravitational potential at the height z.

SOLUTION 4-7
In differential form, (4.17) can be written as

$$d\phi = -G \frac{dm}{r}$$

In this case, we don't need to do a volume integral, because we will just be integrating over the surface. Since it's a disk, we take a basic piece of area to be small ring of width dx. The area of the ring is

$$2\pi x\, dx$$

We can use this to obtain the differential element of mass. Let the surface density of mass be σ. Since we are told the mass is distributed uniformly, this is a constant. Then we have $dm = 2\pi\sigma x\,dx$. Now the distance from an arbitrary point on the disk to a point z on the central axis is

$$r = \sqrt{x^2 + z^2}$$

We get the answer by adding up all of the little disks, that is, by integrating over the entire radius $0 \le x \le R$. We have

$$\phi = -G \int_0^R \frac{2\pi\sigma x}{\sqrt{x^2 + z^2}}\,dx$$

Hence, the potential turns out to be

$$\phi = -2\pi\sigma G\sqrt{x^2 + z^2}\,\Big|_0^R = -2\pi\sigma G\left[\sqrt{R^2 + z^2} - z\right]$$

EXAMPLE 4-8
Find the force on the particle described in Example 4-7.

SOLUTION 4-8
The force is related to potential energy via

$$F = -\nabla V$$

In the case of gravity we saw earlier (4.16) that

$$V = m\phi$$

And so

$$F = -m\nabla\phi$$

For the present example, the gradient is only along the z direction. Some tedious calculation shows that

$$F = -m\frac{d\phi}{dz} = 2\pi m\sigma G\left[\frac{z}{\sqrt{R^2 + z^2}} - 1\right]$$

Quiz

1. What is the acceleration due to gravity near the surface of the sun?

2. A fashion model weighs 120 lb in New York City. How much would she weigh near the surface of the sun?

3. What is the gravitational force between two rocks each weighing 8.5 kg and 0.5 m apart?

4. What is the escape velocity at the surface of the sun?

CHAPTER 5

Moments of Inertia

Moment

Consider a mass m placed on a beam supported by some type of balance that is free to rotate, as shown in Fig. 5-1. The mass is being acted on by the force of gravity which will tend to cause the beam to rotate toward the ground. We place the *fulcrum,* or the point about which the beam will rotate, at the origin. Then we define the *moment M* as the value of the mass multiplied by the distance of the mass from the fulcrum

$$M = mx \tag{5.1}$$

As the system is currently defined, if a mass is placed to the left of the fulcrum, then its moment is negative since $m > 0$ and $x < 0$. Therefore

$$\text{moment left of origin} = -mx$$

Now consider placing two masses m_1 and m_2 on a beam, as shown in Fig. 5-2. Notice that mass m_1, which is situated to the left of the fulcrum, will cause the beam to rotate in the *counter-clockwise* direction. Meanwhile, mass m_2, which

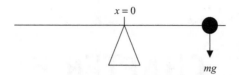

Fig. 5-1 A body placed on a lever arm. The moment is the distance of the body from the origin multiplied by its mass.

is placed to the right of the origin, will cause the beam to rotate in the *clockwise* direction.

Given that the two masses are tending to rotate the beam in different directions, under what conditions will the beam not rotate? We call this condition *balance.* For two masses on a balance beam, the condition for balance is

$$m_1x_1 + m_2x_2 = 0 \qquad (5.2)$$

When there are several masses in the system, we can calculate the total moment of the system by summing up each of the individual moments

$$M_{\text{Total}} = \sum_{i=1}^{n} m_i x_i \qquad (5.3)$$

Hence, if we place *n* masses on the beam, the condition for balance is

$$\sum_{i=1}^{n} m_i x_i = 0 \qquad (5.4)$$

In summary, when the total moment of a system is zero, the masses balance and the beam will not rotate.

EXAMPLE 5-1

Two masses $m_1 = 30$ kg and $m_2 = 50$ kg are placed on a balance beam as shown in Fig. 5-2. The mass m_2 is 2 m to the right of the fulcrum. Where should we place mass m_1 so that the system is balanced?

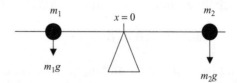

Fig. 5-2 Two masses placed on a balance.

SOLUTION 5-1
For the system to be balanced, (5.2) must be satisfied. We have

$$m_1 x_1 + m_2 x_2 = 0$$

Therefore, mass m_1 should be placed at

$$x_1 = -\frac{m_2}{m_1} x_2 = -\frac{50}{30}(2) = -3.33 \text{ m}$$

Arbitrary Fulcrum and Center of Mass

If the fulcrum is not located at the origin but is instead at some arbitrary position $x = X$, then the condition for balance is

$$\sum_{i=1}^{n} m_i (x_i - X) = 0 \tag{5.5}$$

EXAMPLE 5-2
Three masses, $m_1 = 10$ kg, $m_2 = 7$ kg, and $m_3 = 19$ kg, are placed on a balance beam as shown in Fig. 5-3. The positions of the masses are $x_1 = 0$, $x_2 = 2$, and $x_3 = 11$ respectively, where distances are measured in meters. Where should the fulcrum be placed so that the system is balanced?

SOLUTION 5-2
From (5.5) we have

$$m_1 (x_1 - X) + m_2 (x_2 - X) + m_3 (x_3 - X) = 0$$

With $m_1 = 10$ kg, $m_2 = 7$ kg, and $m_3 = 19$ kg and $x_1 = 0$, $x_2 = 2$, and $x_3 = 11$ this becomes

$$0 = 10(0 - X) + 7(2 - X) + 19(11 - X)$$

$$= -10X + 14 - 7X + 209 - 19X$$

$$= 223 - 36X$$

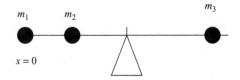

Fig. 5-3   The system described in Example 5-2.

Therefore the fulcrum should be placed at

$$X = \frac{-223}{-36} = 6.2 \text{ m}$$

In Example 5-3, we were really solving for a quantity called the *center of mass*. Let's formalize the definition. Rewriting (5.5), we can isolate the total mass m in the system

$$\sum_{i=1}^{n} m_i (x_i - X) = \sum_{i=1}^{n} m_i x_i - m_i X = \sum_{i=1}^{n} m_i x_i - \sum_{i=1}^{n} m_i X$$

We are free to pull the constant value X outside of the summation to give

$$\sum_{i=1}^{n} m_i x_i - \sum_{i=1}^{n} m_i X = \sum_{i=1}^{n} m_i x_i - X \sum_{i=1}^{n} m_i$$

The sum $\sum_{i=1}^{n} m_i$ is the total mass of the system. Looking back at (5.5) we recall that when the system is in balance, the moment will vanish. Therefore we can write this as

$$\sum_{i=1}^{n} m_i x_i - X \sum_{i=1}^{n} m_i = 0, \Rightarrow$$

$$\sum_{i=1}^{n} m_i x_i = X \sum_{i=1}^{n} m_i$$

Solving for X, which is the center of mass, we find

$$X = \frac{\sum_{i=1}^{n} m_i x_i}{\sum_{i=1}^{n} m_i} \tag{5.6}$$

Therefore to calculate the center of mass for a system consisting of a set of point masses

- Calculate the moment about $x = 0$
- Divide by the total mass in the system.

EXAMPLE 5-3
Four masses $m_1 = 3\,\text{kg}, m_2 = 8\,\text{kg}, m_3 = 1\,\text{kg}$, and $m_4 = 5\,\text{kg}$ are located at the positions $x_1 = 1\,\text{m}, x_2 = 3\,\text{m}, x_3 = 3\,\text{m}$, and $x_4 = 10\,\text{m}$ respectively. Where is the center of mass located for this system?

SOLUTION 5-3
The moment about the origin is given by

$$M = \sum_{i=1}^{n} m_i x_i$$

For the data given in this problem, the moment is

$$M = (3)(1) + (8)(3) + (1)(4) + (5)(10)$$

$$= 3 + 24 + 4 + 50$$

$$= 81$$

Notice that the units of M are kg-m. The total mass of the system is

$$m = \sum_{i=1}^{4} m_i = 3 + 8 + 1 + 5 = 17\,\text{kg}$$

Therefore the center of mass is located at

$$X = \frac{M}{m} = \frac{81\,\text{kg-m}}{17\,\text{m}} = 4.8\,\text{m}$$

Next, we consider the center of mass for a system of point masses in three dimensions.

The center of mass or *centroid* in three dimensions has coordinates $\bar{x}, \bar{y}$, and $\bar{z}$ given by

$$\bar{x} = \frac{\sum_{i=1}^{n} m_i x_i}{\sum_{i=1}^{n} m_i}, \qquad \bar{y} = \frac{\sum_{i=1}^{n} m_i y_i}{\sum_{i=1}^{n} m_i}, \qquad \bar{z} = \frac{\sum_{i=1}^{n} m_i z_i}{\sum_{i=1}^{n} m_i} \qquad (5.7)$$

EXAMPLE 5-4
Three masses, $m_1 = 2, m_2 = 4$, and $m_3 = 6$ where masses are given in kg are located at points $(x,y,z) = (1,-1,2), (1,0,0)$, and $(0,2,0)$ respectively, where position is measured in meters. What are the coordinates of the center of mass?

SOLUTION 5-4

First we calculate the total mass of the system. We obtain

$$\sum_{i=1}^{3} m_i = 2 + 4 + 6 = 12 \text{ kg}$$

Now

$$\sum_{i=1}^{3} m_i x_i = (2)(1) + (4)(1) + (6)(0) = 6$$

$$\sum_{i=1}^{3} m_i y_i = (2)(-1) + (4)(0) + (6)(2) = 10$$

$$\sum_{i=1}^{3} m_i z_i = (2)(2) + (4)(0) + (6)(0) = 4$$

Therefore, the center of mass coordinates are

$$\bar{x} = \frac{6}{12} = \frac{1}{2}$$

$$\bar{y} = \frac{10}{12} = \frac{5}{6}$$

$$\bar{z} = \frac{4}{12} = \frac{1}{3}$$

Continuous Systems

The formulas we have looked at so far are readily generalized to continuous systems. We simply replace the summations by integrals, and so the coordinates of the center of mass become

$$\bar{x} = \frac{\int x \, dm}{\int dm}, \quad \bar{y} = \frac{\int y \, dm}{\int dm}, \quad \bar{z} = \frac{\int z \, dm}{\int dm} \tag{5.8}$$

From the center of mass coordinates, we can define the first moments with respect to the xy, yz, and xz planes. These are

$$Q_{yz} = m\bar{x}, \quad Q_{xz} = m\bar{y}, \quad Q_{xy} = m\bar{z} \tag{5.9}$$

For a region R in the x-y plane, where $a \leq x \leq b$ and $c \leq y \leq d$ we can define the centroid $(\bar{x}, \bar{y})$ as

$$\bar{x} = \frac{\int_a^b x \, dA}{A}, \quad \bar{y} = \frac{\int_c^d y \, dA}{A} \tag{5.10}$$

where A is the area of the region R. The moment of R about the y axis is given by

$$Q_x = \int_a^b x \, dA \tag{5.11}$$

while the moment of R about the x axis is given by

$$Q_y = \int_c^d y \, dA \tag{5.12}$$

EXAMPLE 5-5
Find the first moments for the area of the x-y plane bounded by $y = ax^2$, $y = 0$, and $x = b$.

SOLUTION 5-5
The area in question is illustrated in Fig. 5-4.
    The first moments that we will calculate for this problem are

$$Q_x = \int y \, dA$$

$$Q_y = \int x \, dA$$

    In the first case, the area is calculated from a small horizontal strip with a height given by dy. The area of the strip is then $dA = (b - x)dy$, as illustrated in Fig. 5-5.

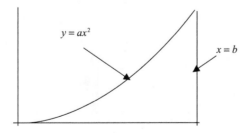

Fig. 5-4   In area used in Example 5-5.

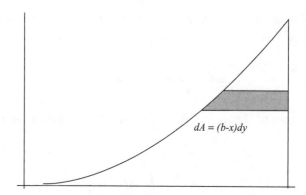

$dA = (b-x)dy$

Fig. 5-5 A horizontal strip of differential area.

With $y = ax^2$, at $x = 0$ we have $y = 0$, while at $x = b$ we have $y = ab^2$. Therefore the first moment is

$$Q_x = \int y\,dA = \int_0^{ab^2} (b-x)y\,dy = \int_0^{ab^2} \left(b - \sqrt{\frac{y}{a}}\right) y\,dy$$

$$= \int_0^{ab^2} by\,dy - \frac{1}{\sqrt{a}} \int_0^{ab^2} y^{3/2}dy = b\frac{y^2}{2} - \frac{2}{5\sqrt{a}} y^{5/2} \Big|_0^{ab^2}$$

$$= \frac{a^2 b^5}{2} - \frac{2a^2 b^5}{5} = \frac{a^2 b^5}{10}$$

To calculate the other moment, we consider a vertical strip of width dx. This is illustrated in Fig. 5-6.

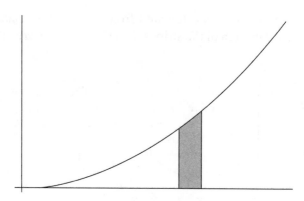

Fig. 5-6 Differential strip of width dx used to calculate $Q_y = \int x\,dA$.

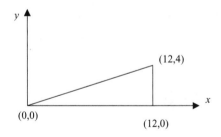

Fig. 5-7   A region R defined by a triangle.

In this case, the differential area is just $dA = ydx = ax^2dx$, with $0 \le x \le b$. Therefore

$$Q_y = \int x\, dA = \int_0^b ax^3dx = a\frac{x^4}{4}\Big|_0^b = \frac{ab^4}{4}$$

EXAMPLE 5-6
Consider the region R defined by a triangle in the x-y plane with vertices at $(0,0)$, $(12,0)$, and $(12,4)$, as shown in Fig. 5-7. Calculate the centroid of this region.

SOLUTION 5-6
The first step is to find a relationship between x and y. This can be done by considering the line that passes through the points $(0,0)$ and $(12,4)$. Since this is just a straight line we find that

$$y = \frac{x}{3}$$

The area of R is the area of the triangle $\frac{1}{2}bh$, which in this case is

$$A = \frac{1}{2}(12)(4) = 24$$

The moment about the y axis is

$$Q_x = \int_0^{12} x\left(\frac{x}{3}\right)dx = \frac{1}{3}\int_0^{12} x^2dx = \frac{x^3}{9}\Big|_0^{12} = 192$$

The x coordinate of the centroid is found by dividing the moment by the area

$$\bar{x} = \frac{Q_x}{A} = \frac{192}{24} = 8$$

Now inverting the relationship on the line that passes through the points (0,0) and (12,4) we have

$$x = 3y$$

And

$$dA = (12 - x)\,dy = (12 - 3y)\,dy$$

Therefore the moment about the x axis is

$$Q_y = \int_0^4 y\,(12 - 3y)\,dy = \int_0^4 12y - 3y^2\,dy = 6y^2 - y^3 \Big|_0^4 = 96 - 64 = 16$$

Hence the y coordinate of the centroid is

$$\bar{y} = \frac{Q_y}{A} = \frac{16}{24} = \frac{2}{3}$$

Second Moments, Moment of Inertia of an Area

The *second moment* or the *axial moment of inertia I* of an element of area dA about an axis in the plane is the product of the area and the square of the distance from the given axis. In the xy plane

$$dI_x = y^2 dA$$
$$dI_y = x^2 dA \tag{5.13}$$

And so the axial moment of inertia, which is calculated by integrating (5.13) is

$$I_x = \int y^2 dA$$

$$I_y = \int x^2 dA \tag{5.14}$$

EXAMPLE 5-7
Determine the axial moment of inertia of a rectangle with base b and height h about the x axis. The rectangle is placed as shown in Fig. 5-8.

SOLUTION 5-7
To determine the differential area, we take a small strip running parallel to the x axis. This is shown in Fig. 5-9.

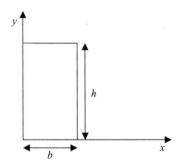

Fig. 5-8　The rectangular region used in Example 5-7.

The area of a little strip parallel to the x axis is

$$dA = bdy$$

We can find the axial moment of inertia by adding up all the little strips from $y = 0$ to $y = h$. Therefore we find

$$I = \int y^2 dA$$

$$= \int_0^h y^2 (bdy)$$

$$= b \int_0^h y^2 dy$$

$$= b\frac{y^3}{3}\bigg|_0^h$$

$$= b\frac{h^3}{3}$$

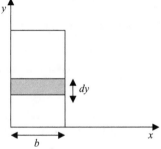

Fig. 5-9　Take a strip of differential area. This is a little rectangle (the size is exaggerated) with area equal to its base times its height, which is *bdy*.

Moment of Inertia for Mass

The moment of inertia for an element of mass about an axis is the product of the mass element dm and the square of the distance of that mass element from the given axis. In general, if that distance is r then

$$I = \int r^2 dm \qquad (5.15)$$

In three dimensions, if we use Cartesian coordinates then the moments of inertia about the three axes are

$$I_x = \int (y^2 + z^2)\, dm \qquad (5.16)$$

$$I_y = \int (x^2 + z^2)\, dm \qquad (5.17)$$

$$I_z = \int (x^2 + y^2)\, dm \qquad (5.18)$$

To consider calculation of moments of inertia in multiple dimensions, we start with a simpler case. Imagine a distribution of mass over some surface S. Let the mass density be σ where

$$\sigma = \frac{\text{total mass}}{\text{surface area}} = \frac{m}{A} \qquad (5.19)$$

The moment of inertia about the z axis is

$$I_z = \int_S (x^2 + y^2)\, \sigma\, dS \qquad (5.20)$$

We make this notion more concrete with an example.

EXAMPLE 5-8
Find the moment of inertia about the z axis for a spherical shell of radius a and total mass m.

SOLUTION 5-8
The surface area of a sphere of radius a is $4\pi a^2$. Therefore, the mass density in this case is

$$\sigma = \frac{m}{4\pi a^2}$$

To do the integral, we will need to use spherical polar coordinates. To integrate over the surface of a sphere, the differential element of surface area is

$$dS = a^2 \sin\theta d\theta d\phi$$

The coordinates x and y are given by

$$x = a\sin\theta\cos\phi, \qquad y = a\sin\theta\sin\phi$$

Therefore we have

$$x^2 + y^2 = a^2\sin^2\theta\cos^2\phi + a^2\sin^2\theta\sin^2\phi = a^2\sin^2\theta$$

Hence, in this case (5.20) becomes

$$
\begin{aligned}
I_z &= \frac{ma^2}{4\pi} \int_0^{2\pi}\int_0^{\pi} \sin^3\theta\, d\theta\, \phi \\
&= \frac{ma^2}{4\pi}(2\pi)\int_0^{\pi} \sin^3\theta\, d\theta \\
&= \frac{ma^2}{2}\int_0^{\pi} \sin\theta\,(1 - \cos^2\theta)\, d\theta \\
&= \frac{ma^2}{2}\int_0^{\pi} \sin\theta\, d\theta - \frac{ma^2}{2}\int_0^{\pi} \sin\theta\,(\cos^2\theta)\, d\theta \\
&= ma^2 - \frac{ma^2}{3} = \frac{2ma^2}{3}
\end{aligned}
$$

EXAMPLE 5-9
Find the moment of inertia about the y axis for a bar of length l and mass m laid along the x axis.

SOLUTION 5-9
Since the bar is laid along the x axis, the distance r of a mass element from the axis is x. Hence we can write (5.15) as

$$I = \int x^2 dm$$

Now a mass element dm is the mass per unit length times an infinitesimal amount of distance which in this case is dx. So

$$dm = \frac{m}{l}dx$$

Therefore the moment of inertia is

$$I = \int_{-l/2}^{l/2} x^2 \left(\frac{m}{l}\right) dx$$

$$= \frac{m}{l} \int_{-l/2}^{l/2} x^2 dx$$

$$= \frac{m}{3l} x^3 \Big|_{-l/2}^{l/2}$$

$$= \frac{m}{3l} \left(\frac{2l^3}{8}\right) = \frac{ml^2}{12}$$

EXAMPLE 5-10
Find the moment of inertia about the x axis for a right circular cylinder of height h and radius a.

SOLUTION 5-10
This time we use the volume density to calculate dm. The volume of a right circular cylinder of radius a and height h is

$$V = \pi a^2 h$$

In spherical coordinates, the differential element of volume is

$$dV = r\, dr\, d\theta\, dz$$

Therefore the differential element of mass is

$$dm = \frac{m}{V} dV = \frac{m}{\pi a^2 h} r\, dr\, d\theta\, dz$$

The cylinder is shown in Fig. 5-10. The moment of inertia can be calculated using

$$I_x = \int (y^2 + z^2)\, dm$$

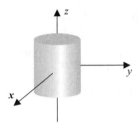

Fig. 5-10   In Example 5-10, we find the moment of inertia of a right circular cylinder.

where $x = r \cos\theta$, $y = r \sin\theta$, and $z = z$ are the relationships between Cartesian and cylindrical coordinates. With these transformations in mind, the moment of inertia is

$$I_x = \frac{m}{\pi a^2 h} \int_{-h/2}^{h/2} \int_0^{2\pi} \int_0^a (r^2 \sin^2\theta + z^2) r\, dr\, d\theta\, dz$$

$$= \frac{m}{\pi a^2 h} \int_{-h/2}^{h/2} \int_0^{2\pi} \int_0^a r^3 \sin^2\theta\, dr\, d\theta\, dz + \frac{m}{\pi a^2 h} \int_{-h/2}^{h/2} \int_0^{2\pi} \int_0^a z^2 r\, dr\, d\theta\, dz$$

$$= \frac{ma^2}{4\pi h} \int_{-h/2}^{h/2} \int_0^{2\pi} \sin^2\theta\, d\theta\, dz + \frac{m}{2\pi h} \int_{-h/2}^{h/2} \int_0^{2\pi} z^2\, d\theta\, dz$$

$$= \frac{ma^2}{4\pi} \int_0^{2\pi} \sin^2\theta\, d\theta + \frac{mh^2}{24\pi} \int_0^{2\pi} d\theta$$

$$= \frac{ma^2}{4} + \frac{mh^2}{12} = \frac{m}{12}(3a^2 + h^2)$$

EXAMPLE 5-11
Find the moment of inertia I_y for a sphere with total mass m and radius a.

SOLUTION 5-11
The volume of a sphere of radius a is

$$V = \frac{4}{3}\pi a^3$$

The differential element of volume is

$$dV = r^2 \sin\theta\, dr\, d\theta\, d\phi$$

Therefore the differential element of mass is

$$dm = \frac{m}{V}dV = \frac{3m}{4\pi a^3}r^2 \sin\theta\, dr\, d\theta\, d\phi$$

Cartesian coordinates are related to spherical polar coordinates via

$$x = r\sin\theta\cos\phi$$
$$y = r\sin\theta\sin\phi$$
$$z = r\cos\theta$$

The integral we need to calculate is

$$I_y = \int (x^2 + z^2)\, dm$$

Now

$$x^2 + z^2 = r^2 \sin^2\theta \cos^2\phi + r^2 \cos^2\theta$$

Therefore the moment of inertia is given by

$$I_y = \frac{3m}{4\pi a^3}\int_0^{2\pi}\int_0^{\pi}\int_0^{a}(r^2\sin^2\theta\cos^2\phi + r^2\cos^2\theta)r^2\sin\theta\, dr\, d\theta\, d\phi$$

Let's do each piece separately. The left piece is

$$\frac{3m}{4\pi a^3}\int_0^{2\pi}\int_0^{\pi}\int_0^{a}(r^2\sin^2\theta\cos^2\phi)r^2\sin\theta\, dr\, d\theta\, d\phi$$

$$= \frac{3m}{4\pi a^3}\int_0^{2\pi}\int_0^{\pi}\int_0^{a}(r^4\sin^3\theta\cos^2\phi)\, dr\, d\theta\, d\phi$$

Integrating over the radial coordinate, this becomes

$$\frac{3ma^2}{20\pi}\int_0^{2\pi}\int_0^{\pi}\sin^3\theta\cos^2\phi\, d\theta\, d\phi$$

Recalling the trig identity

$$\cos^2\phi = \frac{1 + \cos 2\phi}{2}$$

The integral over the ϕ coordinate gives

$$\int_0^{2\pi} \left(\frac{1+\cos 2\phi}{2}\right) d\phi = \frac{\phi}{2} + \frac{\sin 2\phi}{4}\bigg|_0^{2\pi} = \pi$$

So, for the left piece we are left with

$$\frac{3ma^2}{20} \int_0^{\pi} \sin^3 \theta \, d\theta = \frac{3ma^2}{20} \int_0^{\pi} \sin\theta(1-\cos^2 \theta) \, d\theta$$

Now

$$\int_0^{\pi} \sin\theta(1-\cos^2\theta) \, d\theta = \int_0^{\pi} \sin\theta \, d\theta - \int_0^{\pi} \sin\theta \cos^2\theta \, d\theta$$

$$= -\cos\theta\bigg|_0^{\pi} + \frac{\cos^3\theta}{3}\bigg|_0^{\pi} = 2 - \frac{2}{3} = \frac{4}{3}$$

Putting everything together, the left piece becomes

$$\frac{3ma^2}{20} \int_0^{\pi} \sin^3\theta \, d\theta = \frac{3ma^2}{20}\left(\frac{4}{3}\right) = \frac{ma^2}{5}$$

To get the moment of inertia, we need to add the right side of the integral which was

$$\frac{3m}{4\pi a^3} \int_0^{2\pi}\int_0^{\pi}\int_0^{a} (r^2 \cos^2\theta) r^2 \sin\theta \, dr \, d\theta \, d\phi$$

An exercise in tedious calculation shows this also comes out to

$$\frac{3m}{4\pi a^3} \int_0^{2\pi}\int_0^{\pi}\int_0^{a} (r^2 \cos^2\theta) r^2 \sin\theta \, dr \, d\theta \, d\phi = \frac{ma^2}{5}$$

Adding the two pieces gives the moment of inertia for the sphere (see Fig. 5-11)

$$I_y = \frac{2ma^2}{5}$$

Fig. 5-11 In Example 5-11, we calculate the moment of inertia for a sphere of mass m and radius a.

Radius of Gyration

As if the calculation of moment of inertia was not enough torture, we forge ahead. Now we turn to a subject dear to the heart of Elvis, the *radius of gyration*. This quantity is simply calculated by taking the square root of the moment of inertia divided by the mass. Denoted by the letter k, the radius of gyration is

$$k = \sqrt{\frac{I}{m}} \qquad (5.21)$$

More specifically,

$$k_x = \sqrt{\frac{I_x}{m}}$$

gives the distance k_x of the radius of gyration from the x axis, and similarly for I_y and I_z. At the distance given by k, we can view the equivalent mass as a point mass that has an equivalent moment of inertia.

EXAMPLE 5-12
Find the radius of gyration for I_y as calculated for a sphere of mass m and radius a.

SOLUTION 5-12
In Example 5-11, we found that

$$I_y = \frac{2ma^2}{5}$$

So

$$\frac{I_y}{m} = \frac{2a^2}{5}$$

And the radius of gyration for the sphere is

$$k = \sqrt{\frac{2a^2}{5}} = \sqrt{\frac{2}{5}}\, a$$

Parallel Axis Theorem

Let I_{cm} be the moment of inertia about an axis that passes through the center of mass of a body. Then the moment of inertia about a parallel axis that is a distance r from the axis passing through the center of mass is

$$I_r = I_{cm} + mr^2 \tag{5.22}$$

Quiz

1. Two masses $m_1 = 75$ kg and $m_2 = 15$ kg are placed on a balance beam with the fulcrum located at the origin. If the first mass is located at $x = -7$ m, where should the second mass be placed so that the beam is balanced?

2. If masses $m_1 = 7$ kg, $m_2 = 4$ kg, and $m_3 = 8$ kg are located at $x_1 = 2$ m, $x_2 = -2$ m, and $x_3 = 5$ m, find the center of mass of the system.

3. Two masses $m_1 = 10$ kg and $m_2 = 3$ kg are located at $(x, y, z) = (1, 1, 1)$ and $(-1, 2, 3)$ respectively. Where is the center of mass?

4. Find Q_x and Q_y for the region bounded by $y^2 = ax$, $y = 0$, and $x = b$.

5. Find the centroid of a triangle defined by the vertices $(0,0)$, $(5,0)$, and $(5,1)$.

6. Find the moment of inertia I_y for a right circular cylinder of mass m, height h, and radius R.

7. Find the moment of inertia I_z for a sphere of mass m and radius R.

8. Find the moment of inertia I_y for a right circular cone of height h and base of radius R where the axis of the cone is placed along the x axis, and the tip of the cone is at the origin.

9. For the cylinder in problem 6, find the radius of gyration.

10. For the cone in problem 8, find the radius of gyration.

CHAPTER 6

Cables

In this chapter we extend the look we took at statics and equilibrium in Chapter 3 by considering a special case, a brief study of *cables*.

Cables

Consider a *loaded* cable. That is, a cable suspended at both ends that supports some weight such as a bridge. The cable carries a load of f given in N/m or lb/ft and it sags a distance d below the horizontal drawn between the two supports, as shown in Fig. 6-1. The distance covered by the bridge below is called the span, which we denote by s.

The tension at the midpoint of the cable can be found from the load, the sag, and the span of the bridge as follows

$$T_M = \frac{fs^2}{8d} \tag{6.1}$$

Statics and Dynamics Demystified

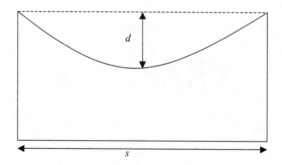

Fig. 6-1 Basic layout of a hanging cable.

EXAMPLE 6-1
A suspension bridge with a span of 140 m is supported by two cables, each of which carries a load of 10220 N/m. If the sag of each cable is 10 m, what is the tension at the midpoint of the cable?

SOLUTION 6-1
The tension at the midpoint can be found from a straightforward application of (6.1)

$$T_M = \frac{fs^2}{8d} = \frac{(10220\,\text{N/m})(140\,\text{m})^2}{8\,(10\,\text{m})} = 2.5\,\text{MN}$$

Endpoint Tension in Cables

The tension T_e in the supports of the cable can be found from the same parameters. In this case

$$T_e = \frac{1}{2}fs\sqrt{1 + \frac{s^2}{16d^2}} \tag{6.2}$$

EXAMPLE 6-2
Each cable in a suspension bridge with a 700-ft span and a 47-ft sag carries a load of 920 lb/ft. What is the tension at the midpoint of the cable and what is the tension at the supports found on each end?

SOLUTION 6-2
The tension at the midpoint, using (6.1) is

$$T_M = \frac{(920\,\text{lb/ft})(700\,\text{ft})^2}{8\,(47\,\text{ft})} = 1,198,936\,\text{lb}$$

The tension at the supports can be found using (6.2) which gives

$$T_e = \frac{1}{2}fs\sqrt{1 + \frac{s^2}{16\,d^2}} = \frac{(920)\,(700)}{2}\sqrt{1 + \frac{(700)^2}{16(47)^2}} = 1{,}241{,}423\,\text{lb}$$

Length of a Suspension Cable

The actual linear length of a suspension cable can be accurately estimated knowing only the span s of the cable and the sag d. This is done with a series expansion. If we denote the length of the cable by l then

$$l = s\left[1 + \frac{8}{3}\left(\frac{d}{s}\right)^2 - \frac{32}{5}\left(\frac{d}{s}\right)^4 + \frac{256}{7}\left(\frac{d}{s}\right)^6 + \cdots\right] \qquad (6.3)$$

Given that the sag is usually much less than the span of the cable, the ratio $\frac{d}{s}$ quickly goes to zero. In most cases

$$l \approx s\left[1 + \frac{8}{3}\left(\frac{d}{s}\right)^2\right]$$

EXAMPLE 6-3
How long is the cable used in Example 6-2?

SOLUTION 6-3
The ratio of the sag to the span is

$$\frac{d}{s} = \frac{47\,\text{ft}}{700\,\text{ft}} = 0.07$$

The square of this term is a measly 0.0045, so terms in the series expansion of l rapidly go to zero with increasing power. Therefore we take

$$l \approx s\left[1 + \frac{8}{3}\left(\frac{d}{s}\right)^2\right]$$

And find that l is about 831 ft long.

Catenary Cables

A *catenary* is a cable that carries the load along the cable instead of horizontally. The curve is described by hyperbolic functions. See your favorite calculus book if you're interested, we're just going to write them down. The basic idea is that they give a parabolic shape like a hanging cable. An example is shown in Fig. 6-2, which is just a plot of the hyperbolic cosine function.

Once again, we assume that the cable is hanging over a span s and that it sags a distance d. The total length of the cable is surprisingly denoted by l, and we continue to denote the load carried by the cable by f. Things are more complicated this time around, so we start with some preliminaries. If the cable carries a load f and the tension at the midpoint of the cable is H, then

$$c = \frac{H}{f} \tag{6.4}$$

With this parameter c in hand, we can relate the x coordinate of a point on the cable, which denotes its position along the span using the usual Cartesian coordinates to the height y of the point using

$$y = c \cosh(x/c) \tag{6.5}$$

The sag and span satisfy

$$c + d = c \cosh(s/2c) \tag{6.6}$$

The tension in the cable at any point y is

$$T = fy \tag{6.7}$$

The total length of the cable is

$$l = 2c \sinh(s/2c) \tag{6.8}$$

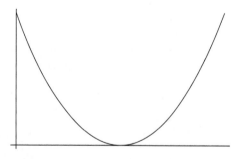

Fig. 6-2 A catenary.

Another relation for the length is

$$(c+d)^2 = c^2 + \frac{l^2}{4} \tag{6.9}$$

The maximum tension in the cable is

$$T_{MAX} = f(c+d) \tag{6.10}$$

EXAMPLE 6-4
A cable carrying a load of 2 lb/ft is suspended between two posts that are 320 ft apart. The sag on the cable is 45 ft. What is the maximum tension in the cable? What is the tension at the midpoint of the cable?

SOLUTION 6-4
To find the maximum tension, we must determine c. This can be done using (6.9)

$$(c+d)^2 = c^2 + \frac{l^2}{4} = c^2 + \frac{(320)^2}{4}$$

$$(c+d)^2 = c^2 + 2cd + d^2$$

Equating both sides

$$2cd + d^2 = \frac{(320\,\text{ft})^2}{4} = 25{,}600\,\text{ft}^2 \Rightarrow$$

$$c = \frac{(25{,}600\,\text{ft}^2 - d^2)}{2d} = \frac{(25{,}600\,\text{ft}^2 - (45\,\text{ft})^2)}{2\,(45\,\text{ft})} = 255\,\text{ft}$$

Then

$$T_{MAX} = f(c+d) = 2\,\text{lb/ft}(255\,\text{ft} + 40\,\text{ft}) = 600\,\text{lb}$$

To find the tension at the midpoint, we apply (6.4)

$$H = fc = (2\,\text{lb/ft})\,(255\,\text{ft}) = 510\,\text{lb}$$

Quiz

1. A suspension bridge is held up by two cables. The span of the bridge is 800 ft and each cable, which sags by 70 ft, carries a horizontal load of 850 lb/ft. What is the tension in the midpoint of the cable?

2. The tension at the endpoint of a suspension bridge with a span of 500 ft is 950,000 lb. If the load carried by each cable is 820 lb/ft, how far down does the center of the support cable sag?

3. A cable sags a distance of 2 ft. The span of the cable is 10 ft. If the cable were taken down and stretched straight out on the ground, how long would it be?

4. A cable carrying a load of 92 N/m is suspended between two towers that are 36 m apart. The sag on the cable is 5 m. What is the maximum tension in the cable? What is the tension at the midpoint of the cable?

CHAPTER 7

Friction

The force of friction is one of the most familiar forces in daily life. Friction is responsible for some of the resistance you feel when trying to start an object moving. For example, consider sliding a book across a wooden table. To start the book sliding, you have to apply some force to get it going. You can sense that something is there resisting your attempt to initiate motion, it might be colloquially described as the roughness of the table surface. Formally, we say that *static friction* between the book and the table is contributing to the resistance of the book when we try to initiate motion.

Once the book starts moving, after a time it begins to slow down and comes to a stop. This behavior is also due to friction, but when a body is in motion we say that there is a force of *kinetic friction* between the body and the surface it is in contact with.

The fact that friction opposes motion tells us something about how we can characterize it mathematically. Since it resists motion, a frictional force will point in the opposite direction to that of an applied force, as shown in Fig. 7-1.

As can be seen in the figure, the frictional force acts tangentially. Let's characterize the two types of frictional forces formally. These are

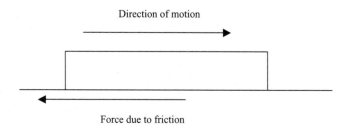

Direction of motion

Force due to friction

Fig. 7-1 The force due to friction points in the opposite direction of the motion of a body.

- Static friction: This type of friction resists the initiation of movement. That is why you feel some resistance when trying to get a book or other object to slide across a table.

- Kinetic friction: This type of friction is the force that opposes the direction of motion and tends to slow down and stop a moving body.

The force due to friction is related to the *normal force* on a body. As its name implies, this force is *normal* or perpendicular to the surface that the body is in contact with. Furthermore, it is proportional to the body's *weight*, which is just the mass of the body times the acceleration due to gravity. If the body is on a flat surface the constant of proportionality is one and the normal force is

$$\mathbf{N} = mg \tag{7.1}$$

This is shown in Fig. 7-2.

The weight of a body always points down toward the center of the earth, but the normal force always points in a direction that is normal to the surface the body is in contact with. Therefore if the body is on a surface which is at an angle with respect to the horizontal, we must resolve the body's weight into components that are normal and tangential to the surface. This can be done using simple trigonometry. As shown in Fig. 7-3, the weight can be resolved into a component that is perpendicular to the surface given by $mg\cos\theta$ and a

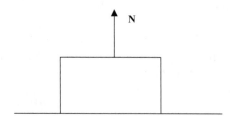

Fig. 7-2 The normal force on a body.

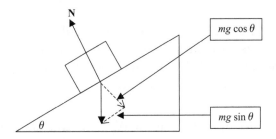

Fig. 7-3   For a body on an inclined surface, we resolve the weight into components that are perpendicular and tangential to the surface. The normal force is equal to the perpendicular component of the weight.

component that is tangential to the surface given by $mg \sin \theta$. The normal force equal to the perpendicular component, that is

$$N = mg \cos \theta \qquad (7.2)$$

Frictional forces are proportional to the normal force, and we call the constant of proportionality the *coefficient of friction*. This coefficient is denoted by the Greek letter μ which is the ratio between the force due to friction and the normal load on the body

$$\mu = \frac{F_{\text{Fric}}}{N} \qquad (7.3)$$

Since it is the ratio of two forces, the coefficient of friction is dimensionless.

There will be a different coefficient of friction for each pair of materials. For example, the friction on your average highway is pretty substantial, i.e., the surface is rough. There will be a different coefficient of friction between your favorite pair of tennis shoes and the surface of the road and the tires of your car and the surface of the road. Now if you visit your nearby ice skating rink, there will be yet again another coefficient of friction between your tennis shoes and the surface of the ice. As you might guess, since the surface of the ice is smooth and slippery, the coefficient of friction between your shoes and the ice is going to be quite a bit smaller than the coefficient of friction between your shoes and the road. This reflects the fact that the frictional forces on the ice which resist motion are quite a bit smaller than those on the road.

A further distinction can be made. As we alluded to earlier, there are two types of friction, static and kinetic. Therefore, we must also specify a *coefficient of static friction* and a *coefficient of kinetic friction* between each pair of materials. Often these are given in a statement of a problem or we can deduce their values by a study of the forces that act on the body.

EXAMPLE 7-1
The coefficient of friction between a flat road and the tires on Bob's sports car is $\mu = 0.62$. If Bob is traveling at 60 miles per hour and hits the brakes, what is his stopping distance in feet? The acceleration due to gravity is $g = 32.2$ ft/s². Assume that the deceleration of the car is constant.

SOLUTION 7-1
We can solve this type of problem using the basic equations of kinematics that you should be familiar with from freshman physics. We will cover these in detail in the next chapter, where we will find that velocity, distance, and acceleration, when the acceleration of the body is constant, are related by (see equation 8.9)

$$v^2 = v_0^2 + 2a(x - x_0) \tag{7.4}$$

Here v is the final velocity of the body, v_0 is the initial velocity of the body, a is its acceleration, and $x - x_0$ is the distance traveled. The initial velocity is given in the problem statement, the first step is to convert it into feet per second

$$(60 \text{ mi/h})(1 \text{ h/3600s})(5280 \text{ ft/mi}) = 88 \text{ ft/s}$$

The car comes to a complete stop, so we take the final velocity $v = 0$. Using (7.4) we find the stopping distance is

$$x - x_0 = -\frac{v_0^2}{2a} \tag{7.5}$$

The road that Bob is driving on is flat, therefore the normal force on Bob's car is just equal to the weight of the car

$$N = mg$$

The force due to friction is proportional to the normal force

$$f = \mu N = \mu mg$$

This is the only force acting on the car. Recalling Newton's second law this is just $F = ma$. Since the frictional force points in the opposite direction to the car's motion, it will be negative. Therefore using $F = ma$ with $F = -f = -\mu mg$ we have

$$-\mu mg = ma, \Rightarrow a = -\mu g$$

Now we use (7.5) to find the stopping distance

$$x - x_0 = -\frac{v_0^2}{2a} = \frac{v_0^2}{2\mu g} = \frac{(88 \text{ ft/s})^2}{2(0.62)(32.2 \text{ ft/s}^2)} = 194 \text{ ft}$$

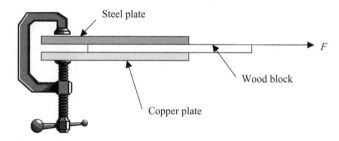

Fig. 7-4   The setup of Example 7-2.

EXAMPLE 7-2
A board is fastened in between two metal plates with a clamp. The top plate
is steel, and the coefficient of friction between steel and the board is 0.3. The
bottom plate is made of copper, and the coefficient of friction between the copper
and the board is 0.27. If the normal force exerted by the clamp is 250 N and a
force F is applied to pull the board out of the assembly, what is the value of F
just before motion impends?

SOLUTION 7-2
The assembly is shown in Fig. 7-4. Since the board is sandwiched in between
two metal plates, we must add up the frictional forces due to each plate when
summing up the forces in the problem. The frictional force is proportional to
the normal force as

$$f_{steel} = \mu_{steel}N$$

$$f_{copper} = \mu_{copper}N$$

Just before motion impends, the system is in equilibrium and we have the sum
of the forces equal to zero $\sum F = 0$. That is

$$F - f_{steel} - f_{copper} = F - \mu_{steel}N - \mu_{copper}N = 0$$

We use negative signs for the frictional forces between the block and each
plate because the frictional force points in an opposite direction to that of the
applied force used to try and pull the block out of the clamp assembly. Solving
for F we find that just before motion impends

$$F = (0.3)\,250 + (0.27)\,250 = 143\,\text{N}$$

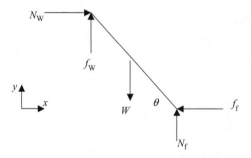

Fig. 7-5 A free body diagram for the ladder in Example 7-3.

EXAMPLE 7-3
A ladder that weighs 27 lb is resting against a wall such that the angle between the ladder and the floor is $\theta = 53°$. What is the coefficient of friction if the ladder is 8 ft long?

SOLUTION 7-3
Since the ladder is at rest, it is in equilibrium. First we draw a free body diagram for the ladder. This is shown in Fig. 7-5.

As shown in the figure, there are normal and frictional forces acting on the ladder at the points of contact with the wall and with the floor. We denote the normal and frictional forces acting at the point of contact with the wall by N_W and f_W and the normal and frictional forces acting at the point of contact with the floor by N_f and f_f. Remember, normal forces act in a direction that is perpendicular to the surface the body is in contact with, and frictional forces act in a direction that is tangential to the surface the body is in contact with.

Since the body is in equilibrium, this tells us that the sum of the forces and moments in each direction will vanish. We use the usual x-y Cartesian coordinate system. Looking at Fig. 7-5, the sum of the forces in the x direction are given by

$$\sum F_x = N_W - f_f = 0, \Rightarrow N_W = f_f \qquad (7.6)$$

Now we sum the moments about the point of contact with the wall. Again, since the ladder is in equilibrium, the sum must vanish. We denote the length of the ladder by l

$$\sum M_W = -\frac{Wl}{2} \cos\theta + N_f l \cos\theta - \mu N_f l \sin\theta = 0 \qquad (7.7)$$

Finally, the sum of the forces in the y direction also vanish since the system is in equilibrium

$$\sum F_y = N_f + \mu N_W - W = 0 \qquad (7.8)$$

Substitution of $N_W = f_f$ into (7.8) allows us to write

$$W = N_f + \mu N_W = N_f + \mu f_f$$

Recalling that a frictional force is proportional to the normal force according to $f = \mu N$ this becomes

$$W = N_f + \mu^2 N_f, \Rightarrow N_f = \frac{W}{1 + \mu^2}$$

Now we substitute for N_f in the equation for the moments (7.7). This gives

$$\sum M_W = 0 = -\frac{Wl}{2} \cos\theta + N_f l \cos\theta - \mu N_f l \sin\theta$$

$$= -\frac{Wl}{2} \cos\theta + \frac{W}{1 + \mu^2} l \cos\theta - \mu \frac{W}{1 + \mu^2} l \sin\theta$$

Rearranging terms we have

$$\frac{2}{Wl \cos\theta} = \frac{1 + \mu^2}{Wl \cos\theta - \mu Wl \sin\theta}$$

Cross multiplying by $Wl \cos\theta - \mu Wl \sin\theta$ and doing some algebra we obtain an equation for the coefficient of friction

$$\mu^2 + 2\mu \tan\theta - 1 = 0$$

Notice that the length of the ladder has canceled. We can solve this by using the quadratic formula

$$\mu = \frac{-b \pm \sqrt{b^2 - 4ac}}{2a} = \frac{-2 \tan\theta \pm \sqrt{4 \tan^2\theta + 4}}{2}$$

Now,

$$2 \tan\theta = 2 \tan 53° = 2.7$$

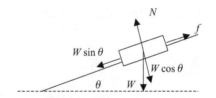

Fig. 7-6 Forces on the brick in Example 7-4.

So we find that

$$\mu = \frac{-2.7 \pm 3.4}{2}$$

We discard the root with the minus sign because coefficients of friction are positive. Therefore we find that $\mu = 0.35$.

EXAMPLE 7-4
A brick is resting on a plastic surface. The coefficient of static friction between the brick and the surface is $\mu = 0.35$. If the surface is tilted, at what angle θ will the brick begin to slide down the surface?

SOLUTION 7-4
The forces involved in this problem are shown in Fig. 7-6. The forces acting on the brick are the normal force, the weight of the brick, and the frictional force.
   Summing the forces in each direction, which will vanish in equilibrium, the normal force is

$$N = W \cos \theta = mg \cos \theta$$

Along the tangential direction, the forces are

$$f - W \sin \theta = 0, \Rightarrow f = W \sin \theta = mg \sin \theta$$

Using the fact that the frictional force is proportional to the normal force via $f = \mu N$, we arrive at the following equation

$$mg \sin \theta = \mu mg \cos \theta$$

Rearranging terms we find

$$\mu = \tan \theta$$

Given that $\mu = 0.35$, the brick will begin to slide when the surface is at an angle θ with respect to the horizontal which is

$$\theta = \tan^{-1} \mu = \tan^{-1} 0.35 = 19°$$

EXAMPLE 7-5
A block is sliding down an incline plane. The acceleration of the block is $a = 5.41 \text{ m/s}^2$ and the plane is at an angle $\theta = 51°$ with respect to the horizontal. What is the coefficient of kinetic friction?

SOLUTION 7-5
The forces on the block are the same as those shown in Fig. 7-6. Therefore once again the normal force is

$$N = W \cos \theta = mg \cos \theta$$

The frictional force is given in terms of the normal force as

$$f = \mu_k N = \mu_k mg \cos \theta$$

Since the block is in motion, the sum of the forces on the block will be equal to mass times acceleration according to Newton's second law

$$\sum F = ma$$

Besides friction, the only other force acting on the block is the component of the weight along the direction of the motion. This force points in a downward direction, in the same direction as the motion of the block. Therefore we take this as the positive direction. The frictional force points in the opposite way to the direction of motion, so we take it to be negative. Putting everything together $\sum F = ma$ gives us

$$mg \sin \theta - \mu_k mg \cos \theta = ma$$

where $mg \sin \theta$ is the component of the block's weight along its direction of motion. Dividing through by the mass and rearranging terms we find that the coefficient of kinetic friction is

$$\mu_k = \frac{g \sin \theta - a}{g \cos \theta} = \frac{(9.81 \text{ m/s}^2) \sin 51° - 5.41 \text{ m/s}^2}{(9.81 \text{ m/s}^2) \cos 51°} = 0.36$$

Belt Friction

You have probably spent a great deal of time solving problems involving a rope, cord, or belt looped around a pulley. In such cases it might be desirable to find the tensions in each end of the rope, say.

We now consider a belt looped over a pulley, but this time we consider the case where the pulley and belt are rough so that there is a coefficient of friction

Statics and Dynamics Demystified

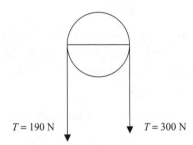

$T = 190$ N $T = 300$ N

Fig. 7-7   Diagram representing a belt wrapped around a pulley as described in
Example 7-6.

between them. A question of interest is what are the tensions in both ends of
the belt when slippage is about to occur. The tensions on each side of the pulley
are related by a nice little formula

$$T_1 = T_2 e^{\mu\alpha} \tag{7.9}$$

where $T_1 > T_2$ and α is the angle of wrap of the belt around the pulley in radians.
This formula only holds when a slip is about to occur.

EXAMPLE 7-6
A belt is wrapped around a pulley as shown in Fig. 7-7. On the slack side, the
tension is 190 N while the tension in the tight side is 300 N. Find the coefficient
of friction between the belt and the pulley.

SOLUTION 7-6
The situation is shown in Fig. 7-7. Looking at (7.9), we have

$$T_1 = 300\,\text{N}, \ T_2 = 190\,\text{N}$$

Looking at the figure, we see that the belt is wrapped around the diameter of
the pulley, in other words

$$\alpha = 180° = \pi\,\text{radians}$$

Let's rewrite (7.9) so that we can solve the coefficient of friction. Rearranging
terms we have

$$\frac{T_1}{T_2} = e^{\mu\alpha}$$

Solving for the coefficient of friction we obtain

$$\mu = \frac{1}{\alpha} \ln\left(\frac{T_1}{T_2}\right)$$

Putting in the numbers given in the problem, we find that

$$\mu = \frac{1}{\pi} \ln \left(\frac{300}{190} \right) = 0.15$$

Rolling Resistance

At one time or another, you might have gotten the wheels of your car stuck in a depression in a dirt road. To get out, you had to apply just the right amount of gas so that the wheels of the car didn't slip but instead just managed to climb out of the depression. This situation is a case of *rolling resistance*. For a wheel with radius r we illustrate the situation in Fig. 7-8.

When motion is about to occur and the wheel will get out of the depression, we call the distance a indicated in Fig. 7-8 the *rolling resistance*. It is related to the load W on the wheel, the radius r of the wheel, and the applied force F in the following way

$$a = \frac{F}{W} r \qquad\qquad (7.10)$$

Notice that the coefficient of rolling resistance is a distance.

EXAMPLE 7-7
A cart has wheels that are 13 inches in diameter. A wheel carrying a load of 500 lb is stuck in a depression, and it is found that a force of 300 lb is required to roll the cart out. What is the coefficient of rolling resistance?

SOLUTION 7-7
Applying (7.10) we find

$$a = \frac{(300)(6.5)}{500} = 3.9\,\text{inches}$$

EXAMPLE 7-8
On a test rail track at a top-secret facility, which cannot be named, a rail car has a wheel with a 700 mm diameter that carries a 750 N load. The coefficient of

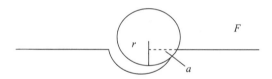

Fig. 7-8   Rolling resistance.

rolling resistance between the wheel and the track is 0.41 mm. What force is required such that the wheel will begin to roll on the track?

SOLUTION 7-8
Rearranging terms in (7.10), we have

$$F = \frac{Wa}{r} = \frac{(750)(0.41)}{(350)} = 0.88\,\text{N}$$

Quiz

1. A block that weighs 210 lb is being pulled with a force F on a flat surface. Just before motion ensues, what is the force being applied to the block if the coefficient of static friction between the block and the floor is 0.38?

2. A block is secured between two copper plates with a clamp. The coefficient of friction between the block and copper is $\mu = 0.28$. If a force is applied to pull the block out of the assembly, what is the strength of that force if the normal force exerted by the clamp is 113 N?

3. A block is sliding down a surface which is at an angle $\theta = 67°$ with the horizontal. If the coefficient of kinetic friction is $\mu_k = 0.25$, what is the acceleration of the block in SI units?

4. A belt is wrapped around a pulley such that the tension on the tight side of the belt is 150 N and the angle of wrap is $5\pi/3$. The coefficient of friction between the belt and the pulley is 0.32. Find the tension in the slack side of the belt just before the belt slips.

5. If the coefficient of rolling resistance is 0.29 mm, what force must be applied to set a wheel rolling on a track if it carries a 3000 N load and the diameter of the wheel is 800 mm?

CHAPTER 8

Particle Kinematics

We now turn our attention to the study of motion. A full study of motion, which relates the forces on a particle to its acceleration via Newton's second law, forms the basis of the science of *dynamics*. In this chapter, however, we are going to start with a simple analysis of some one- and two-dimensional cases of motion that does not explicitly involve forces. This type of study of motion is known as *kinematics*. We begin with the simplest type of motion possible, motion along a straight line.

Rectilinear Particle Motion

Let's consider motion along a straight line. We denote the position along this line by x. If a particle is moving along a straight line, the first question we can ask is how fast is the particle moving. Speed or velocity (we will be more precise about the distinction between speed and velocity later) is nothing but distance traveled in a given time interval. So if a particle moves between points x_1 and x_2 between times t_1 and t_2, then the average velocity is given by

$$\bar{v} = \frac{x_2 - x_1}{t_2 - t_1} \tag{8.1}$$

As we will see in a moment, we will use the derivative to write down a precise definition of velocity. We can see how to approach this definition by considering the average velocity in the following way. That is, the *average velocity* of a particle moving from the point x to the point $x + \Delta x$ in the time interval t to $t + \Delta t$, then the average velocity can be written as

$$\bar{v} = \frac{x + \Delta x - x}{t + \Delta t - t} = \frac{\Delta x}{\Delta t} \tag{8.2}$$

With this definition in hand, we are ready to consider the notion of *instantaneous velocity*. To obtain the instantaneous velocity, we let the time interval shrink to zero. That is

$$v = \lim_{\Delta t \to 0} \frac{\Delta x}{\Delta t} = \frac{dx}{dt} \tag{8.3}$$

Velocity is position per time, so we might measure velocity in ft/s or if we are using SI units, m/s where distance is measured in meters. Now we have the tools in hand to characterize the particle's motion in terms of position and velocity. We can go a step further—as you might have guessed there is one important piece of data missing—the *acceleration* of the particle. Acceleration tells us how rapidly the velocity of the particle is changing. Following the procedure used to define velocity, we can define the average acceleration and the instantaneous acceleration. The average velocity is given in terms of the velocities v_1 and v_2 measured at times t_1 and t_2 as

$$\bar{a} = \frac{v_2 - v_1}{t_2 - t_1} \tag{8.4}$$

Or, a better definition of average acceleration is

$$\bar{a} = \frac{\Delta v}{\Delta t} \tag{8.5}$$

Now let's take the limit as $\Delta t \to 0$. This gives us the instantaneous acceleration

$$a = \lim_{\Delta t \to 0} \frac{\Delta v}{\Delta t} = \frac{dv}{dt} \tag{8.6}$$

Now, since acceleration is the time rate of change of velocity and velocity is the time rate of change of position, we can write acceleration in terms of position by considering the second derivative. That is

$$a = \frac{d^2x}{dt^2} \qquad (8.7)$$

From (8.6) we can guess at the units. There is one dimension of time in the denominator, while the numerator has dimensions of distance/time. Therefore acceleration has dimensions of distance/time squared. In U.S. units, when we are measuring distance in feet and time in seconds, we measure acceleration in ft/s^2. In SI units acceleration is given in meters per second squared or m/s^2.

Constant Acceleration

We have already created a simple situation by restricting ourselves to motion along one dimension. Now we're going to keep things easy by going one more step and only considering the special case of constant acceleration. Let x_0 and v_0 denote the initial position and velocity of the particle, respectively. Then the following equations hold so long as the acceleration a is constant

$$v = v_0 + at \qquad (8.8)$$

$$v^2 = v_0^2 + 2a\,(x - x_0) \qquad (8.9)$$

$$x - x_0 = v_0 t + \frac{1}{2}at^2 \qquad (8.10)$$

$$x - x_0 = \frac{1}{2}\,(v + v_0)\,t \qquad (8.11)$$

These equations can be derived using Newton's laws, but it's not necessary to worry about that right now. We will just accept them as given and use them to solve kinematics problems. The trick to solving these types of problems is to look at what information you're given, what information is missing, and what information you want to find out. That simply means being careful and judicious about picking the right equations from (8.8) to (8.11) to work with.

EXAMPLE 8-1
A rocket sled is moving along a straight track. Its position is known to increase with time according to $x\,(t) = 3t^2 - 2t$ where time is measured in seconds and position is measured in feet. Find the velocity and acceleration of the sled at $t = 5$ s.

SOLUTION 8-1

We use (8.3) to find the velocity of the sled and (8.6) to find the acceleration. First, we compute the derivative of $x(t)$ to obtain the functional form of the velocity

$$v(t) = \frac{dx}{dt} = \frac{d}{dt}(3t^2 - 2t) = 6t - 2$$

At $t = 5$ s, the velocity is $v(t) = 6(5) - 2 = 28$ ft/s. To find the acceleration, we take the derivative of $v(t)$

$$a = \frac{dv}{dt} = \frac{d}{dt}(6t - 2) = 6 \text{ ft/s}^2$$

EXAMPLE 8-2

A red sports car driven by a balding middle-aged man is speeding down a rural highway. Fearing the impending presence of a police officer, he breaks to slow from 90 miles per hour to 45 miles per hour over a distance of 300 ft. How much time does it take for the car to slow to 45 miles per hour? Find the acceleration assuming that it's constant.

SOLUTION 8-2

Since the acceleration is constant, we can use (8.8)–(8.11). We're asked to find the time required for the car to slow down, but it's likely this will only take seconds but the velocities are given in miles per hour. First let's convert the velocities to feet per second

$$v_0 = \left(90\frac{\text{miles}}{\text{hour}}\right)\left(\frac{1 \text{ hour}}{3600 \text{ s}}\right)\left(\frac{5280 \text{ ft}}{\text{mile}}\right) = 132 \text{ ft/s}$$

$$v = \left(45\frac{\text{miles}}{\text{hour}}\right)\left(\frac{1 \text{ hour}}{3600 \text{ s}}\right)\left(\frac{5280 \text{ ft}}{\text{mile}}\right) = 66 \text{ ft/s}$$

Now, how can we find the time required to slow down, given only the velocities and the breaking distance? Looking at the equations for constant acceleration, it appears that (8.11) will do the job. We find

$$t = \frac{2(x - x_0)}{v + v_0} = \frac{2(300 \text{ ft})}{(132 + 66) \text{ ft/s}} = 3 \text{ s}$$

To find the acceleration, we use (8.9). The acceleration is

$$a = \frac{v^2 - v_0^2}{2(x - x_0)} = \frac{(132 \text{ ft/s})^2 - (66 \text{ ft/s})^2}{2(300 \text{ ft})} = 21.8 \text{ ft/s}^2$$

EXAMPLE 8-3
A woman is speeding down a straight road in her BMW to get away from her angry husband. Starting from rest and traveling with constant acceleration, she passes her mother-in-law's house traveling at 90 miles per hour. A mile later the car is going 130 miles per hour. What is her acceleration? How long did it take her to travel from rest to her mother-in-law's house? How long did it take her to go the final mile?

SOLUTION 8-3
To find the acceleration, first let's write down what we know. We are told that the car starts from rest, therefore, we know that from the starting point to the mother in laws house

$$v_0 = 0$$

We also know that the car is traveling at 90 miles per hour at that point. Looking at (8.8)–(8.11), however, it seems we don't have quite enough information to solve the problem. A possible equation we can use is (8.9), since we only need to know two velocities and the distance traveled between them. We are given this information in the last part of the problem. So let's take the mother-in-law's house as the starting point in which case we take the initial velocity to be 90 miles per hour

$$v_0 = 90 \text{ mph} = 132 \text{ ft/s}$$

At the end of the mile, the velocity is 130 miles per hour, so we have

$$v = 130 \text{ mph} = 191 \text{ ft/s}$$

The distance traveled between the points at which these speeds are measured is 1 mile, therefore

$$x - x_0 = 5280 \text{ ft}$$

Using (8.9) and solving for the acceleration, we find

$$a = \frac{v^2 - v_0^2}{2(x - x_0)} = \frac{(191 \text{ ft/s})^2 - (132 \text{ ft/s})^2}{2(5280 \text{ ft})} = 1.8 \text{ ft/s}^2$$

With the acceleration in hand, we have enough information to find out the other quantities. The time required to travel from rest to the mother-in-law's house can be found using (8.8)

$$t = \frac{v - v_0}{a} = \frac{132 \text{ ft/s} - 0 \text{ ft/s}}{1.8 \text{ ft/s}^2} = 73 \text{ s}$$

Using the same equation, we find that the time required to go the last mile is

$$t = \frac{v - v_0}{a} = \frac{191 \text{ ft/s} - 132 \text{ ft/s}}{1.8 \text{ ft/s}^2} = 33 \text{ s}$$

Near the surface of the earth, the acceleration due to gravity is constant. Therefore we can apply the equations (8.8)–(8.11) to problems involving falling objects by setting the acceleration equal to the acceleration due to gravity. We denote this acceleration by g. In U.S. units, $g = 32.2 \text{ ft/s}^2$ and in SI units $g = 9.81 \text{ m/s}^2$. Typically, we denote the vertical or up-down direction by the y coordinate. Now remember that the acceleration in the case of free fall points down–so we will have to modify the equations to reflect this. We can accomplish this by putting a negative sign in front of each term involving acceleration. Therefore the equations of free fall in a gravitational field are

$$v = v_0 - gt \tag{8.12}$$

$$y - y_0 = v_0 t - \frac{1}{2} g t^2 \tag{8.13}$$

$$v^2 - v_0^2 = -2g(y - y_0) \tag{8.14}$$

$$y - y_0 = \frac{1}{2}(v + v_0)t \tag{8.15}$$

$$y - y_0 = vt + \frac{1}{2} g t^2 \tag{8.16}$$

EXAMPLE 8-4

A man in a hot air balloon that is 95 m high and is rising with a velocity of 2.2 m/s drops a baseball out of the balloon basket. How long does it take the baseball to hit the ground, and what is its velocity at impact?

SOLUTION 8-4

First let's write down what we know. At the time of release, the baseball is traveling at the same velocity as the balloon. The balloon is rising at a velocity of 2.2 m/s, therefore the initial velocity of the baseball is

$$v_0 = +2.2 \text{ m/s}$$

The initial height $y_0 = 95$ m. We take the ground to be the origin and therefore $y = 0$ at impact. Let's find the velocity at impact first using (8.14). We have

$$v = -\sqrt{v_0^2 - 2g(y - y_0)} = -\sqrt{(2.2\,\text{m/s})^2 - 2(9.81\,\text{m/s}^2)(0 - 95)\,\text{m}}$$

$$= -\sqrt{4.84\,\text{m}^2/\text{s}^2 + 1864\,\text{m}^2/\text{s}^2} = -\sqrt{1868.7\,\text{m}^2/\text{s}^2} = -43 \text{ m/s}$$

We take the negative square root because the ball is heading downward—toward negative y, so its velocity must be negative. With this piece of information in hand, we can solve for the time to impact using (8.15)

$$t = \frac{2(y - y_0)}{(v + v_0)} = \frac{2(0 - 95)\,\text{m}}{(-43 \text{ m/s} + 2.2 \text{ m/s})} = 4.7 \text{ s}$$

EXAMPLE 8-5
A tourist on a lookout tower drops a camera. It strikes the ground with a speed $v = 174$ ft/s. How high is the lookout tower? How long did it take for the camera to reach the ground?

SOLUTION 8-5
We are given three pieces of information in the problem. The first is obvious—the speed $v = 174$ ft/s the camera has when it reaches the ground. We are also told the tourist simply drops the camera—therefore we take the initial speed to be $v_0 = 0$. Finally we know that since the camera is traveling straight down—it is moving with constant acceleration g.

Looking at the equations of free fall (8.12)–(8.16), it appears that (8.14) is the best one to use in order to find the distance that the camera travels. If we take the upward direction to be positive y, then the distance $y - y_0$ will be negative. We will call it $-h$ and then

$$-h = \frac{v^2 - v_0^2}{-2g}$$

Now, since we are taking downward to be negative, we need to write the velocity as $v = -174$ ft/s because the velocity points down. So we obtain

$$h = \frac{v_0^2 - v^2}{-2g} = \frac{0 - (-174\,\text{ft/s})^2}{-2(32.2\,\text{ft/s}^2)} = 470 \text{ ft}$$

The time required for the camera to reach the ground can be found using (8.12). We find

$$t = \frac{v - v_0}{-g} = \frac{-174 \text{ ft/s}}{-32.2 \text{ ft/s}^2} = 5.4 \text{ s}$$

Particle Motion in Higher Dimensions

The next step in the basic analysis of particle motion is to consider motion in two and three dimensions. This will require the introduction of vectorial quantities. We will denote the unit vectors in the x, y, and z directions by the bold-face letters $\mathbf{i}$, $\mathbf{j}$, and $\mathbf{k}$ respectively. Imagine a particle is moving on some curve in three-dimensional space. Then the position vector $\mathbf{r}$ that points from the origin to the position of the particle is simply

$$\mathbf{r} = x\mathbf{i} + y\mathbf{j} + z\mathbf{k} \tag{8.17}$$

Derived quantities like velocity and acceleration are calculated in an analogous manner to that used in the one-dimensional case. Velocity is the time rate of change of position and so the velocity vector is given by

$$\mathbf{v} = \frac{dx}{dt}\mathbf{i} + \frac{dy}{dt}\mathbf{j} + \frac{dz}{dt}\mathbf{k} \tag{8.18}$$

At this point we will make a distinction between speed and velocity. Strictly speaking velocity is a vector with magnitude and direction, while speed is simply the magnitude of the velocity vector. Speed is given by

$$v = |\mathbf{v}| = \sqrt{\dot{x}^2 + \dot{y}^2 + \dot{z}^2} \tag{8.19}$$

where we have used the dot as a shorthand notation for a derivative with respect to time.

The acceleration vector is the time rate of change of velocity. We will write it in terms of the second derivative of position

$$\mathbf{a} = \frac{d\mathbf{v}}{dt} = \frac{d^2\mathbf{r}}{dt^2} = \ddot{x}\mathbf{i} + \ddot{y}\mathbf{j} + \ddot{z}\mathbf{k} \tag{8.20}$$

Again, the two dots are shorthand for a second derivative with respect to time, i.e.,

$$\ddot{x} = \frac{d^2x}{dt^2} \; .$$

In two-dimensional motion where we take $\mathbf{r} = x\,\mathbf{i} + y\,\mathbf{j}$, the velocity vector is tangent to the curve, which represents the path of the particle. Such a path is a curve $y(x)$. The angle between the velocity vector and the x axis is

$$\theta = \tan^{-1}\left(\frac{\dot{y}}{\dot{x}}\right) = \tan^{-1}\left(\frac{dy}{dx}\right) \tag{8.21}$$

Note that the acceleration vector is not necessarily tangent to the path of the particle.

Sometimes it is helpful to analyze particle motion in terms of unit vectors that are tangent and normal to the curve describing the particles path. Since these vectors are orthogonal and form a basis, it is possible to actually write the velocity and acceleration vectors in terms of these new unit vectors instead of using $\mathbf{i}$, $\mathbf{j}$, $\mathbf{k}$. We already know that the velocity vector $\mathbf{v}$ is tangent to the curve, so this allows us to write down a unit vector tangent to the curve immediately. This can be done by dividing the velocity vector (8.18) by its magnitude or by the speed (8.19). If we denote the unit tangent vector to the curve by $\mathbf{T}$ then

$$\mathbf{T} = \frac{\dot{x}\,\mathbf{i} + \dot{y}\,\mathbf{j} + \dot{z}\,\mathbf{k}}{\sqrt{\dot{x}^2 + \dot{y}^2 + \dot{z}^2}} \tag{8.22}$$

Now consider a curve that is parameterized by s such that $\mathbf{r} = \mathbf{r}(s)$. Then it turns out that $\mathbf{T} = \dfrac{d\mathbf{r}}{ds}$. For an arbitrary curve, we can define the curvature k as

$$k = \left|\frac{d\mathbf{T}}{ds}\right| = \frac{|d\mathbf{T}/dt|}{ds/dt}$$

If we invert this quantity we obtain the radius of the curve

$$\rho = \frac{1}{k} = \frac{1}{|d\mathbf{T}/ds|}$$

This allows us to write the acceleration vector in terms of unit vectors that are tangential and normal to the curve

$$\mathbf{a} = \frac{d^2s}{dt^2}\mathbf{T} + \frac{|v|^2}{\rho}\mathbf{N} = \frac{d^2s}{dt^2}\mathbf{T} + \left|\frac{ds}{dt}\right|^2 k\mathbf{N} \tag{8.23}$$

The unit vector that is normal to the curve can be calculated easily by taking the time derivative of **T** and dividing by its magnitude

$$N = \frac{d\mathbf{T}/dt}{|d\mathbf{T}/dt|} \tag{8.24}$$

EXAMPLE 8-6
Suppose that a particle is on a path described by the curve $x(t) = t^2$, $y(t) = 3t^3 + 2t$, $z(t) = 7 + 3t$. Find the particle's velocity and a unit vector tangent to the curve.

SOLUTION 8-6
The velocity vector is

$$\mathbf{v} = \frac{d\mathbf{r}}{dt} = \dot{x}\,\mathbf{i} + \dot{y}\,\mathbf{j} + \dot{z}\,\mathbf{k} = 2t\,\mathbf{i} + (9t^2 + 2)\mathbf{j} + 3\,\mathbf{k}$$

The magnitude of this vector is

$$|\mathbf{v}| = \sqrt{\dot{x}^2 + \dot{y}^2 + \dot{z}^2} = \sqrt{(2t)^2 + (9t^2 + 2)^2 + (3)^3} = \sqrt{81t^4 + 40t^2 + 13}$$

To obtain the unit tangent vector, we divide the velocity vector by its magnitude

$$\mathbf{T} = \frac{\mathbf{v}}{|\mathbf{v}|} = \frac{2t\mathbf{i} + (9t^2 + 2)\mathbf{j} + 3\,\mathbf{k}}{\sqrt{81t^4 + 40t^2 + 13}}$$

EXAMPLE 8-7
A particle is constrained to move in the x-y plane such that

$$x(t) = A\cos\omega t$$
$$y(t) = A\sin\omega t$$

What is the speed of the particle? What is the acceleration of this particle? Find unit vectors that are tangent and normal to the curve that describes the particle's path.

SOLUTION 8-7
First, we write down the position vector

$$\mathbf{r} = A\cos\omega t\,\mathbf{i} + A\sin\omega t\,\mathbf{j}$$

The velocity vector is the time rate of change of the position vector

$$\mathbf{v} = \frac{d\mathbf{r}}{dt} = -\omega A \sin \omega t \, \mathbf{i} + \omega A \cos \omega t \, \mathbf{j}$$

The speed is just the magnitude of this vector

$$v = |\mathbf{v}| = \sqrt{\omega^2 A^2 \sin^2 \omega t + \omega^2 A^2 \cos^2 \omega t} = \omega A$$

The acceleration vector is the time rate of change of velocity

$$\mathbf{a} = \frac{d\mathbf{v}}{dt} = \frac{d}{dt}(-\omega A \sin \omega t \, \mathbf{i} + \omega A \cos \omega t \, \mathbf{j}) = -\omega^2 A \cos \omega t \, \mathbf{i} - \omega^2 A \sin \omega t \, \mathbf{j}$$

Notice that $\mathbf{a} = \omega^2 \mathbf{r}$, a characteristic of uniform circular motion that we will see in the next section. The magnitude of this vector is

$$a = |\mathbf{a}| = \sqrt{\omega^4 A^2 \cos^2 \omega t + \omega^4 A^2 \sin^2 \omega t} = \omega^2 A$$

We can construct the unit tangent vector using (8.22)

$$\mathbf{T} = \frac{\mathbf{v}}{|\mathbf{v}|} = -\sin \omega t \, \mathbf{i} + \cos \omega t \, \mathbf{j}$$

Next we construct the unit normal vector using (8.24). First we have

$$\frac{d\mathbf{T}}{dt} = -\omega \cos \omega t \, \mathbf{i} - \omega \sin \omega t \, \mathbf{j}$$

The magnitude of this vector is

$$\left| \frac{d\mathbf{T}}{dt} \right| = \sqrt{(-\omega \cos \omega t)^2 + (-\omega \sin \omega t)^2} = \omega$$

So, the normal vector is

$$\mathbf{N} = \frac{\dfrac{d\mathbf{T}}{dt}}{\left| \dfrac{d\mathbf{T}}{dt} \right|} = -\cos \omega t \, \mathbf{i} - \sin \omega t \, \mathbf{j}$$

Projectile Motion

As a trivial special case of two-dimensional motion, we consider the case of a projectile being fired at some muzzle velocity v_0. Acted upon by gravity, it lands some distance R from the point where it was fired.

The first step in analyzing the kinematics of projectile motion is to resolve the initial or *muzzle* velocity into x and y components. We imagine that the gun or cannon is elevated at some angle θ with respect to the horizon. We can then use basic trigonometry to find the desired quantities (see Fig. 8-1).

The motion of the projectile in the x and y directions is completely independent. That is, we can analyze the kinematics of the motion with respect to x and y directions by considering the motion in x along using what we already know about one-dimensional motion and then considering the motion in y alone using what we know about one-dimensional free fall. The common denominator is the muzzle velocity.

Let's start then by considering the horizontal motion of the projectile. In the x direction, when we stick to the simplifying case of no air resistance, there are no forces acting and the projectile has a constant initial velocity. Now since acceleration is the time rate of change of velocity—there is no acceleration in this case since the derivative of a constant vanishes. When is it valid to consider a case of no air resistance? A general rule of thumb is that the muzzle velocity is small.

When firing a projectile there are two pieces of data that will be of interest when considering the horizontal motion. These are the time of flight and the total distance that the projectile will travel, or it's *range*. If we consider the case with $a = 0$, looking at the equations of one-dimensional kinematics (8.8)–(8.11), we see that there is only one equation that is of interest to us. Equation (8.10)—which includes time and distance traveled—is the equation we seek. Let's restate it here

$$x - x_0 = v_0 t + \frac{1}{2}at^2$$

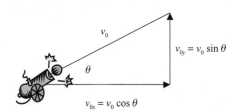

Fig. 8-1 The components of muzzle velocity.

If we set the acceleration to zero it becomes

$$x - x_0 = v_0 t$$

Now we make the substitution $v_0 \to v_{0x}$ as described in Fig. 8-1, giving us the equation we need

$$x - x_0 = v_0 \cos \theta \, t \qquad (8.25)$$

Next, we consider the motion of the projectile in the vertical or y direction. The motion in this direction is governed by one force—gravity. The force of gravity controls its motion over the entire trajectory and is what brings the projectile back to earth. As we saw earlier, the force of gravity makes its way into the equations of kinematics via the constant g in equations (8.12)–(8.16). Since there is acceleration in the y direction, this means that the y component of velocity must be changing with time. So we set $v_0 \to v_{0y} = v_0 \sin \theta$. When considering projectile motion, the following equations are of use

$$v_y = v_0 \sin \theta - gt \qquad (8.26)$$

$$y - y_0 = v_0 \sin \theta \, t - \frac{1}{2} g t^2 \qquad (8.27)$$

$$v_y^2 = v_0^2 \sin^2 \theta - 2g(y - y_0) \qquad (8.28)$$

The velocity in the y direction goes to zero at the top of the trajectory. At this point, the velocity is changing direction. On the upward part of the curve the velocity points upward. At the very top of the curve it momentarily goes to zero, and then on the downward part of the curve the velocity vector is pointing down. Therefore to find the time at which the projectile reaches maximum height, we set $v_y = 0$ in (8.26). Then we find the time to maximum height is

$$t = \frac{v_0 \sin \theta}{g} \qquad (8.29)$$

Next, we want to find out what that maximum height that the projectile reaches is. We can do this by substituting this value of t into (8.27) and solving for $y - y_0$.

A final item to consider is the total time of flight. The height of the projectile is zero at launch and at impact. Therefore to find the time to impact we can set $y - y_0$ to zero in (8.27) and solve. We obtain the following relation

$$0 = \left(v_0 \sin \theta - \frac{1}{2} g t \right) t \qquad (8.30)$$

The multiplicative factor on the outside corresponds to the time of launch, $t = 0$. Setting the term in parentheses equal to zero and solving we find the time to impact is

$$t_{\text{impact}} = \frac{2v_0 \sin \theta}{g} \tag{8.31}$$

Of course this is exactly twice the time required for the projectile to reach its maximum height (8.29).

Finally, we want to find the range or total horizontal distance traveled by the projectile. The range can be found by inserting the time to impact (8.31) into (8.25). If we set $R = x - x_0$ then

$$R = v_0 \cos \theta \; t_{\text{impact}} = v_0 \cos \theta \left(\frac{2v_0 \sin \theta}{g} \right) = \frac{v_0^2}{g} \sin 2\theta \tag{8.32}$$

To obtain this result, we have used the relation $\sin 2\theta = 2 \sin \theta \cos \theta$. Finally, we want to find the angle at which the range is maximum. Using simple calculus we have

$$\frac{dR}{d\theta} = \frac{2v_0^2}{g} \cos 2\theta = 0$$

Therefore, the condition for the range to be a maximum is $\cos 2\theta = 0$. For $0 \le \theta \le 90°$, this is true when

$$\theta = \theta_{\text{max}} = 45° \tag{8.33}$$

EXAMPLE 8-8
An artillery piece is situated atop a mountain that is 2.4 km above the valley below. If the gun is situated at an angle of $\theta = 32°$ above the horizontal, what is the time of flight if a fired projectile reaches a target 6.7 km distant?

SOLUTION 8-8
We are told that the target is 6.7 km distant, therefore

$$x - x_0 = 6700 \text{ m}$$

Further we know the time of flight can be obtained from (8.25). Solving for time we can rewrite this equation as

$$t = \frac{x - x_0}{v_0 \cos \theta}$$

Houston we have a problem—while we have found an equation for the time of flight, it contains another unknown quantity. The problem statement has not given us the muzzle velocity v_0. But we do know the height of the mountain. We can use our expression for time together with equation (8.27). This will be of use if we set $y - y_0 = -2400$ m, the height of the mountain. We have

$$y - y_0 = v_0 \sin\theta\, t - \frac{1}{2}gt^2$$

$$= v_0 \sin\theta \left(\frac{x - x_0}{v_0 \cos\theta}\right) - \frac{1}{2}g\left(\frac{x - x_0}{v_0 \cos\theta}\right)^2$$

A bit of algebra allows us to solve for the muzzle velocity

$$v_0 = \frac{x - x_0/\cos\theta}{\sqrt{\dfrac{2(y - y_0)}{g} + \dfrac{2\sin\theta}{g\cos\theta}(x - x_0)}}$$

Putting in the parameters given in the problem, we find that $v_0 = 215$ m/s. Then, the time of flight is

$$t = \frac{x - x_0}{v_0 \cos\theta} = \frac{6700}{(215)\cos 32°} = 36.8 \text{ s}$$

Uniform Circular Motion

If the acceleration of a particle has constant magnitude but its direction is changing, and the velocity of the particle is constant, the particle will travel in a circular path. Such a particle will "feel" an acceleration pointing toward the center of that circular path, think of riding on a merry-go-round. The acceleration that points toward the center is called the *centripetal acceleration* and is related to the velocity of the particle and the radius of the circular path via

$$a = \frac{v^2}{r} \tag{8.34}$$

EXAMPLE 8-9
A car traveling round a circular track with a radius of 200 feet rounds a curve traveling at 45 mph. What is the acceleration of the car?

SOLUTION 8-9

First we write the velocity in feet per second

$$v = 45 \text{ mph} = 66 \text{ ft/sec}$$

Now, we can use (8.34) to find the acceleration

$$a = \frac{v^2}{r} = \frac{(66 \text{ ft/s})^2}{200 \text{ ft}} = 22 \text{ ft/s}^2$$

EXAMPLE 8-10

A satellite is in orbit about the earth at a height $h = 190$ km above the earth's surface. If the orbital speed of the satellite is 7.9 km/s, what is the value of the gravitational constant g at this orbit?

SOLUTION 8-10

The radius in this problem is given by the sum of the height of the satellite above the surface of the earth added to the radius of the earth itself. Then we find that

$$g = \frac{v^2}{r} = \frac{(7900 \text{ m/s})^2}{(6.37 \times 10^6 \text{ m} + 190,000 \text{ m})} = 9.6 \text{ m/s}^2$$

Quiz

1. The position of a particle in meters is given by $x(t) = 2t^3 - 8t$. What are the position, velocity, and acceleration of the particle at $t = 3$ s?

2. The position of an oscillating particle is given by $x(t) = 3e^{-t} \cos 2t$. What is its acceleration as a function of time?

3. A car traveling at 50 miles per hour decelerates at 15 ft/s² before coming to a complete stop. What distance did the car travel during this time?

4. For the car in problem 3, how long did it take to come to a complete stop?

5. A baseball is thrown straight up at 75 miles per hour. How high does it go?

6. How long does it take the baseball to hit the ground?

7. Returning to Example 8-3, what distance does the car cover when traveling from rest to the mother-in-law's house?

8. A man overlooking a cliff drops his keys. They hit the ground traveling at 20 m/s. How long did it take the keys to hit the ground? How high is the cliff?

9. A particle is on a curve parameterized by $x = \cos 2t$, $y = \sin 3t$, $z = t^2 + 7$. Find a unit vector tangent to this curve.

10. Suppose that the motion of a particle is described by $x = t \cos t$, $y = t \sin t$, $z = t^2$. Find the velocity, acceleration, and speed of the particle. Write down the unit vector tangent to the curve.

11. By considering the equations for projectile motion, find an equation for the trajectory $y(x)$.

CHAPTER 9

Dynamics and Newton's Second Law

The science of dynamics describes the change of the motion of a system with time. Mathematically the heart of dynamics is described by Newton's second law, which relates the change in the linear momentum of a system to the forces being impressed upon it. First, we begin with a precise definition of momentum.

Linear Momentum

The linear momentum of a particle is the product of mass m and velocity **v**. Typically momentum is denoted with the symbol **P** however some engineering texts denote momentum by **G**. Since momentum is the product of a scalar (the mass) and a vector (velocity), momentum is a vector quantity. In this book we will use **P** to denote momentum. Therefore we can write

$$\vec{P} = m\vec{v} \tag{9.1}$$

In SI units, we measure mass in kilograms and velocity in meters per second. Therefore the units of momentum, which we denote by $[\vec{P}]$ are

$$[\vec{P}] = [m][\vec{v}] = \text{kg (m/s)} \tag{9.2}$$

In the U.S. system mass is measured in slugs and velocity in feet per second, so the units of momentum are slugs-feet per second.

When a system is isolated, we say that the total momentum is *conserved*. This means that the total momentum in the system is a constant and does not change in time. We will see below that this result is easily derived from Newton's second law. In equation form, we write

$$\vec{P} = \text{a constant (for an isolated system)} \tag{9.3}$$

Conservation of momentum can be used to solve many problems. Although in reality the system in consideration is not *truly* isolated, for practical purposes we can consider it to be so. Determining when it is appropriate to do this requires some simple judgment and experience. For example, if the values of certain quantities you are solving for are much larger than external influences, you can for practical purposes neglect the external influences and solve the problem as if the system were isolated. Another case where you can apply this approximation is the time over which a system is analyzed. In other words, if the external influences do not change much over a short time interval over which you are examining the system, then you can also consider this type of system isolated.

EXAMPLE 9-1
A bullet with a mass of 5 g is fired horizontally from a gun with a speed $v = 1350$ m/s at a wooden block with a mass $M = 9$ kg resting in direct line of sight to the gun on a smooth surface. Assuming the surface is smooth enough so that friction can be neglected, determine the final velocity of the block.

SOLUTION 9-1
We can solve this problem using conservation of momentum. First we calculate the momentum of the bullet. First, notice that the mass of the bullet is given in grams while the mass of the block is given in kilograms. We need to use the same units for the bullet and the block, let's stick with grams. Then applying (9.1), we find that the momentum of the bullet is

$$\vec{P}_i = mv = (5 \text{ g})(1350 \text{ m/s}) = 6750 \text{ g-m/s} \tag{9.4}$$

We have denoted the momentum of the bullet as $\vec{P}_i$ because this is the *initial* momentum in the system. The final momentum of the system will be the momentum of the block after the bullet strikes it. The bullet will strike the block and come to rest inside it, transferring all of its momentum to the block. We are ignoring other factors here, such as the possibility that some heat is transferred to the block when the bullet strikes it. Our assumption will be that the magnitudes of these quantities will be much smaller than those we are considering in the problem. So we simply pretend that they don't exist.

What conservation of momentum tells us via equation (9.3) is that the final momentum of the system will be equal to the initial momentum of the system. That is

$$\vec{P}_i = \vec{P}_f$$

where $\vec{P}_f$ is the final momentum, that is, the momentum of the block. The momentum of the block is given by

$$\vec{P}_f = M\vec{V}$$

where $\vec{V}$ is the block's velocity. Putting everything together conservation of momentum tells us that

$$\vec{P}_i = \vec{P}_f, \Rightarrow m\vec{v} = M\vec{V}$$

Solving for the velocity of the block, we have

$$\vec{V} = \frac{m}{M}\vec{v} = \frac{6750 \text{ g-m/s}}{9000 \text{ g}} = 0.75 \text{ m/s}$$

To obtain the final answer, we used (9.4) and converted the mass of the block from kg to grams.

Newton's Laws of Motion

The foundations for the science of dynamics were laid down in 1666 by Isaac Newton in the English countryside. Newton was only 23 at the time, and he had been forced to leave Cambridge to escape an outbreak of the plague. He returned to the family farm where his mother lived and spent his time contemplating several outstanding problems in physics. It was during this time that the great genius invented calculus so that he could solve problems that had been vexing others for sometime. Among his other discoveries were the *three laws of motion*, which we detail here.

These laws of motion are often framed in terms of "particles." What you consider a "particle" depends on the situation. For example, if you toss a baseball toward home plate it makes sense to think of the baseball as a particle. In another case, consider the earth moving around the sun. If you were a sadist you could try and calculate the force of gravity on every rock, tree, or building on the surface of the earth. We could also try to calculate the motion of the moving baseball, which we've tossed toward home plate with respect to the sun. Or you could simply consider the fact that with respect to the sun, everything on the earth moves as if it were all lumped together as a single body. Over the scale of the solar system, it doesn't matter much, in fact it doesn't matter at all that the baseball has moved a few feet across the surface of the earth when we're considering the sun at 93 million miles away. The baseball is basically a fixed part of the earth. In fact on that scale the earth itself is just a tiny dot. So with respect to the sun, we can consider the earth to be a "particle."

With that in mind, we can consider the three laws of motion.

The First Law of Motion

The first law of motion tells us what happens to a particle that is at rest or in a state of *uniform motion*. By uniform motion, we mean that the velocity of the particle is constant in time, that is,

$$\frac{d\vec{v}}{dt} = 0$$

for a particle in uniform motion. Newton's first law tells us that if a particle is in a state of rest, it will remain in a state of rest if no external forces act upon it. If the particle is in a state of uniform motion, it will remain in a state of uniform motion if no forces act upon it.

The Second Law of Motion

We have used the term *force* without giving a formal definition. We understand what force is intuitively, and Newton's second law gives us a precise mathematical expression that relates force to momentum. Newton's second law tells us that the time rate of change of the momentum of a particle is related to the force acting on the particle

$$\vec{F} = \frac{d\vec{P}}{dt} \tag{9.5}$$

Using the definition of momentum (9.1), we see that Newton's second law can be written as

$$\vec{F} = \frac{d(m\vec{v})}{dt}$$

In many cases, the mass of the particle is a constant and we can pull it outside of the derivative. If the mass is constant then

$$\vec{F} = m\frac{d\vec{v}}{dt}$$

However you should be aware that this is not always the case. A familiar example of a "particle" where the mass is not constant is a rocket. The rocket is burning fuel and ejecting mass out at the back end—therefore the mass of the rocket-fuel system, which we are considering as a particle, is changing with time.

Returning to the constant mass case, recall that acceleration is the time derivative of velocity, that is

$$\vec{a} = \frac{d\vec{v}}{dt}$$

This allows us to write the most familiar form of Newton's law

$$\vec{F} = m\vec{a}$$

Remember, force is a vector. In U.S. units force is measured in pounds (lb), while mass is given in slugs and acceleration in feet per second squared, i.e., ft/s². In SI units acceleration is measured in meters per second squared, m/s² and force is given in *Newtons*, where a Newton is defined using $\vec{F} = m\vec{a}$ as

$$1\,\text{N} = 1\,\frac{\text{kg-m}}{\text{s}^2}$$

We will have a chance to work with Newton's second law in some upcoming example problems.

Historically, scientists and engineers have referred to two different types of mass. These are *inertial* and *gravitational* mass, respectively.

Inertial mass comes into play when determining how a body responds to the application of a given force. More specifically, the acceleration of a particle or body when a given force is applied is determined by the inertial mass via the relation $F = ma$.

In classical mechanics, Newton's second law determines the weight of a particle. In this case the relation $F = ma$ becomes $W = mg$ where W is the weight of the particle or body and g is the acceleration due to gravity. In this case m is the particle's gravitational mass. This mass determines how other bodies or particles will respond to the gravitational force exerted by this particle, in addition to determining how the particle will respond to the gravitational field of another body.

As you might expect, it has been shown through careful experimentation that these two masses are equivalent. A detailed discussion of this issue is beyond the scope of this text, interested parties should consult the references at the back of this book.

Now we state the third law of motion.

The Third Law of Motion

For a given force there is an equal and opposite force. This is the famous saying for every action there is an equal and opposite reaction. Numerically, if a particle exerts a force on a second particle that we denote $\vec{F}_{12}$, the second particle will exert a force $\vec{F}_{21}$ back on the first particle that is equal in magnitude but opposite in direction. That is,

$$\vec{F}_{12} = -\vec{F}_{21} \tag{9.6}$$

In many cases, when doing basic dynamics problems involving forces, multiple forces will be involved. We then write Newton's second law as

$$\sum_i F_i = ma \tag{9.7}$$

In other words the sum of the forces on a particle equals mass times acceleration. We illustrate this with an example, where there is an externally applied force and the force of gravity acting together on an object.

EXAMPLE 9-2
A 3 lb block is being pulled up a plane inclined at an angle $\theta = 25°$ with a force F_{ext} directed along the incline. Assuming that there is no friction, find
   (a) The mass of the block.
   (b) Find the normal force on the block and the minimal force that would have to be applied to keep the block from sliding downward.
   (c) The acceleration of the block when $F_{ext} = 3.3$ lb.

SOLUTION 9-2
   (a) We are told that a 3 lb block is being pulled up the plane. This is the weight of the block. The mass is related to the weight using

$$W = mg$$

Statics and Dynamics Demystified

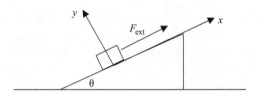

Fig. 9-1   The sliding block in example 9-2.

where g is the acceleration due to gravity. Since the problem is given using U.S. units (lb) we set $g = 32.2$ ft/s^2 and find the mass in slugs, because the units of $W = mg$ are given by

$$[\text{lbs}] = [\text{slugs}][\text{ft/s}^2]$$

We find the mass to be:

$$m = W/g = (3)/(32.2) \text{ slugs} = 0.09 \text{ slugs}$$

(b) To tackle the next part of the problem, we begin by drawing a diagram of the situation and defining a coordinate system. This is shown in Fig. 9-1.

   The simplest way to set up the problem is to orient the coordinate axes so that they correspond to the incline of the plane. Next we draw a free body diagram, which means we show the particle (in this case the box) together with all the forces on it. This is shown in Fig. 9-2.

   To determine the force components in the x and y directions, we use basic trigonometry. Remember that $\cos\theta = $ adj/hyp and $\sin\theta = $ opp/hyp. Now we sum up the forces in a given direction and apply Newton's second

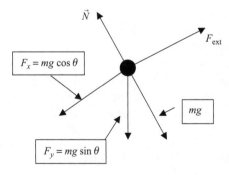

Fig. 9-2   We represent the block as a "particle" and draw a free body diagram, showing all the forces acting on it.

law (9.7). The upward force on the block is the *normal force* $\vec{N}$, which pushed outward, directed from the plane onto the block. If the normal force was greater than gravity the block would fly up off the plane. However, since no external forces are present, it just balances the component of the gravitational force in this direction, which we have taken to be along the y axis. When working with components, we can drop the vector notation since the components of each vector are just scalars. With the block staying affixed to the plane the acceleration in that direction must vanish, so (9.7) gives

$$\sum F_y = N - W \sin\theta = 0$$

Solving we find that the normal force on the block is

$$N = W \sin\theta = (3 \text{ lb}) \sin(25°) = (3 \text{ lb})(0.42) = 1.3 \text{ lb}$$

The minimal force that will keep the block from sliding downward is that for which the block has no acceleration along the x direction. First, looking at Figs. 9-1 and 9-2, we note that the sum of the forces in the x direction is given by

$$\sum F_x = F_{ext} - W \cos\theta$$

Using Newton's second law, if the acceleration in this direction is zero then

$$F_{ext} - W \cos\theta = 0, \quad \Rightarrow F_{ext} = W \cos\theta = (3 \text{ lb}) \cos(25°)$$
$$= (3 \text{ lb})(0.91) = 2.72 \text{ lb}$$

We conclude that if a force that is smaller than 2.72 lb is applied, the block will slide downward.

(c) In this case we have the block accelerating upward, since the force is greater than the minimum force required to keep the block from sliding downward, i.e., $F_{ext} = 3.3 \text{ lb} > 2.72 \text{ lb}$. Therefore we write Newton's second law as

$$\sum F_x = ma_x$$
$$\Rightarrow F_{ext} - W \cos\theta = ma_x$$

Now using the relation between mass and weight the right side is

$$ma_x = \left(\frac{W}{g}\right) a_x$$

On the left side we obtain

$$\sum F_x = F_{ext} - W \cos\theta = 3.3 \text{ lb} - (3 \text{ lb}) \cos 25°$$
$$= 3.3 \text{ lb} - 2.72 \text{ lb} = 0.6 \text{ lb}$$

Then the acceleration is

$$a_x = \frac{\sum F_x}{m} = \left(\frac{g}{W}\right)\sum F_x = \left(\frac{32.2 \text{ ft/s}^2}{3 \text{ lb}}\right)(0.6 \text{ lb}) = 6.44 \text{ ft/s}^2$$

Since Newton's second law is expressed in terms of acceleration, and acceleration is the derivative with respect to time of velocity and velocity is the derivative with respect to time of position, we can express Newton's second law in terms of a second order differential equation that allows us to solve for the position of a particle as a function of time. For simplicity, let's consider one-dimensional motion measured along a coordinate that we surprisingly designate by x. Then the velocity is written as

$$v = \frac{dx}{dt}$$

As we stated above, acceleration is related to velocity by writing

$$\vec{a} = \frac{d\vec{v}}{dt}$$

In terms of position, acceleration is given by

$$a = \frac{d^2x}{dt^2} \qquad (9.8)$$

For the moment, we assume that the mass in the problem is constant. This allows us to write Newton's second law in the alternative form

$$\sum F = m\frac{d^2x}{dt^2} \qquad (9.9)$$

In some texts derivatives with respect to time are denoted with two dots placed above the variable, and so (9.9) can be written as

$$\sum F = m\ddot{x} \qquad (9.10)$$

where it is understood that $\ddot{x} = d^2x/dt^2$.

EXAMPLE 9-3
A block slides down a plane inclined at an angle $\theta = 35°$ with respect to the horizontal. If there is no friction, find the position of the block along the plane, the velocity of the block, and the acceleration of the block all as functions of time and their values at $t = 0.2$ s. Assume that the initial velocity of the block is zero and that the block starts at the top of the plane. Use SI units in this problem.

SOLUTION 9-3
As shown in Fig. 9-3, we place the origin of our coordinate system at the top of the plane.

As in the previous example, there are two forces acting on the block, the gravitational force and the normal force. Using $\sin\theta =$ side opposite/hypotenuse and $\cos\theta =$ side adjacent/hypotenuse, the x and y components of the gravitational force are given by

$$F_x = F_g \sin\theta$$
$$F_y = -F_g \cos\theta$$

In the y direction, the normal force and the y components of the gravitational force exactly cancel—that is once again we assume that the block stays on the inclined plane as it slides down. This means that summing the forces in the y direction gives

$$F_y + N = 0, \quad \Rightarrow N = F_g \cos\theta$$

Now in the x direction, we have

$$\sum F_x = m\ddot{x}$$

In this direction the only force that acts is the x component of the gravitational force, that is $F_x = F_g \sin\theta$, and so we have the following second order

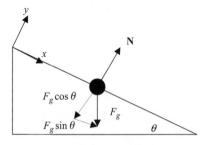

Fig. 9-3   A representation of our coordinate system and forces on a block sliding down an inclined plane. The position of the block at some arbitrary time is represented by the black dot.

differential equation

$$m\ddot{x} = F_g \sin\theta = mg \sin\theta$$

where we used the definition of weight to write the gravitational force as mg. Canceling the mass gives us

$$\ddot{x} = g \sin\theta$$

Therefore the acceleration, which is the second derivative of position with respect to time, is a constant. With $\theta = 35°$ we find that

$$a = g \sin 35° = (9.81 \text{ m/s}^2) \sin 35° = 5.6 \text{ m/s}^2$$

For the velocity, we have

$$\frac{dv}{dt} = g \sin\theta, \quad \Rightarrow \int_0^t dv = \int_0^t g \sin\theta \, d\tau$$

We have chosen τ as a dummy integration variable. Integrating both sides gives us

$$v(t) = g \sin\theta t + v(0) = g \sin\theta t$$

We made the last step because the problem statement says the block starts from rest. At $t = 0.2$ s, the velocity is

$$v(t) = (9.81 \text{ m/s}^2) \sin(35°)(0.2 \text{ s}) = 1.12 \text{ m/s}$$

Finally, we can integrate once more to obtain the position as a function of time. We have

$$\frac{dx}{dt} = g \sin\theta t, \quad \Rightarrow x(t) = \frac{g \sin\theta}{2} t^2$$

Remember the block starts at the top of the plane, where we have put the origin of our coordinate system. Therefore $x(0) = 0$. At $t = 0.2$ s, we find the position to be

$$x = \frac{(9.81 \text{ m/s}^2) \sin(35°)}{2} (0.2 \text{ s})^2 = 0.112 \text{ m}$$

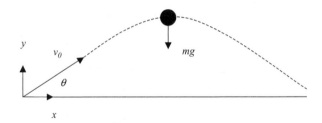

Fig. 9-4   A projectile is analyzed with no air resistance.

EXAMPLE 9-4

A projectile of mass m is fired with an initial velocity $v_0 = 17$ m/s and at an angle $\theta = 37°$. Find the particle's speed as a function of time, its maximum range and the time required for the particle to strike the ground. Neglect air resistance.

SOLUTION 9-4

A schematic of the problem is shown in Fig. 9-4.

When there is no air resistance, the only force that acts on the projectile is the force of gravity. First let's resolve the initial velocity into x and y components. This is easy to do using basic trigonometry as shown in Fig. 9-5.

We can solve the problem by analyzing the motion in each direction independently. With the coordinate system shown in Fig. 9-4, gravity acts downward (i.e., negative) in the y direction. Since there is no force in the x direction, Newton's second law gives

$$\sum F_x = 0 = m\,\frac{d^2x}{dt^2} \tag{9.11}$$

We can cancel the mass term since the equation is just equal to zero and mass is a constant. Integration will give us the velocity in the x direction, which is a constant

$$\frac{dx}{dt} = v_{x_0} = v_0 \cos\theta \tag{9.12}$$

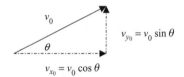

Fig. 9-5   We resolve the initial velocity into x and y components.

Integrating a second time, we obtain the x coordinate as a function of time. We have placed the origin of the coordinate system at the initial position of the particle, therefore we can throw away the constant of integration, giving

$$x(t) = v_{x_0} t = v_0 \cos \theta t \tag{9.13}$$

Now, in the y direction, things are a little more complicated since we need to take into account the gravitational force on the projectile. In this case

$$\sum F_y = -mg = m \frac{d^2 y}{dt^2} \tag{9.14}$$

The minus sign is present because the gravitational force points downward relative to the coordinate system we have chosen. Noting that the mass terms cancel, integration gives us the velocity in the y direction

$$\frac{dy}{dt} = -gt + v_{y_0} = -gt + v_0 \sin \theta \tag{9.15}$$

Integration a second time gives us the y position of the projectile as a function of time. Again, the initial position of the projectile is at the origin, so we can throw away the constant of integration in this case. So we find that

$$y(t) = -\frac{gt^2}{2} + v_0 \sin \theta t \tag{9.16}$$

The speed of the particle is the magnitude of its velocity vector. The x and y components of the velocity vector are given by (9.12) and (9.15), respectively. Therefore the speed is

$$v(t) = \sqrt{\left(\frac{dx}{dt}\right)^2 + \left(\frac{dy}{dt}\right)^2} = \sqrt{v_0^2 \cos^2 \theta + v_0^2 \sin^2 \theta + g^2 t^2}$$

$$= \sqrt{v_0^2 + g^2 t^2} \tag{9.17}$$

To find the time when the projectile strikes the ground, se can simply solve (9.16) when $y(t) = 0$. We have

$$0 = -\frac{g t^2}{2} + v_0 \sin \theta t = t \left(-\frac{gt}{2} + v_0 \sin \theta \right)$$

The first solution $t = 0$ tells us the time when the projectile was launched. Solving for the second time we set

$$-\frac{gt}{2} + v_0 \sin \theta = 0$$

which tells us that the projectile will strike the ground at

$$t = \frac{2v_0 \sin \theta}{g} = \frac{2\,(17\,\text{m/s}) \sin 37°}{9.81\,\text{m/s}^2} \approx 2.1\,\text{s}$$

The range is the value of $x(t)$ when the projectile strikes the ground. Using (9.13) we find

$$x = v_0 \cos \theta \left(\frac{2v_0 \sin \theta}{g} \right) = \frac{v_0^2}{g} \sin 2\theta = \frac{(17\,\text{m/s})^2}{9.81\,\text{m/s}^2} \sin 74° \approx 28.4\,\text{m}$$

In the next example, we consider air resistance.

EXAMPLE 9-5
A projectile is fired straight down from a height h with an initial velocity s. If air resistance is directly proportional to the projectile's velocity, find the position and velocity of the projectile as a function of time.

SOLUTION 9-5
A schematic of the situation is shown in Fig. 9-6.

When we say that air resistance is proportional to the projectile's velocity, this means that the functional form of the air resistance is to multiply the velocity by

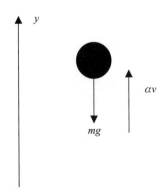

Fig. 9-6   Falling projectile with air resistance.

a constant. If we call the constant of proportionality α then we can write the air resistance as αv. Air resistance for a projectile that is falling in the y direction is a positive upward force. The sum of the forces in the y direction is then

$$\sum F_y = -mg - \alpha v \tag{9.18}$$

Applying Newton's second law we find that

$$m\ddot{y} = -mg - \alpha v \tag{9.19}$$

Since the right hand side is a function of the projectile's velocity, let's use the relation $\ddot{y} = dv/dt$ to write this as an equation involving only the velocity

$$\frac{dv}{dt} = -g - \frac{\alpha}{m}v \tag{9.20}$$

where we have divided through by m. Now we rearrange terms giving

$$\frac{dv}{g + \dfrac{\alpha}{m}v} = -dt$$

Integration gives

$$\ln\left(g + \frac{\alpha}{m}v\right) = -t + C$$

Now we exponentiate both sides

$$g + \frac{\alpha}{m}v = Ce^{-t}$$

This allows us to solve for the velocity as a function of time

$$v(t) = \left(\frac{mC}{\alpha}\right)e^{-t} - \frac{gm}{\alpha} \tag{9.21}$$

Setting $t = 0$ and using the initial velocity $v(0) = s$ allows us to solve for the constant of integration. We have

$$v(0) = s = \left(\frac{mC}{\alpha}\right) - \frac{gm}{\alpha}, \quad \Rightarrow C = \frac{\alpha s}{m} + g$$

So we can write the velocity as

$$v(t) = \left(s + \frac{mg}{\alpha}\right)e^{-t} - g\frac{m}{\alpha} \qquad (9.22)$$

The terminal velocity occurs as $t \to \infty$. In (9.22) the first term vanishes under these conditions, so the terminal velocity is just $-g\frac{m}{\alpha}$. The negative sign indicates that the velocity is pointing down, as we might expect. Now we can use (9.22) to solve for the projectiles position as a function of time. We have

$$v(t) = \frac{dy}{dt} = \left(s + \frac{mg}{\alpha}\right)e^{-t} - g\frac{m}{\alpha}$$

Integrating between $t = 0$ and t we obtain

$$y(t) = \left(s + \frac{mg}{\alpha}\right)(1 - e^{-t}) - g\frac{m}{\alpha}t + h$$

Dynamics and Kinetic Friction

As we described in Chapter 7, kinetic friction is a tangential force between a body in motion with respect to some surface. This force is proportional to the normal force on the body via the relation

$$F_{\text{kin}} = \mu_{\text{k}} N \qquad (9.23)$$

where the constant of proportionality μ_{k} is called the coefficient of kinetic friction. The direction of this force is opposite to that of the direction of motion of the body. For example, if a box is being pushed across a floor to the right, the kinetic friction force points to the left.

EXAMPLE 9-6
A box is placed on a conveyer belt inclined at $15°$ with respect to the horizontal. The conveyer belt is moving with a velocity of 2.44 m/s and the box is initially at rest. If the coefficient of kinetic friction between the box and the conveyer belt is $\mu_{\text{k}} = 5/16$, find the time required for the box to come to rest on the conveyer belt.

SOLUTION 9-6
The scenario is depicted in Fig. 9-7. When the box is placed on the conveyer belt, it will slip backward until the force of kinetic friction stops its motion.

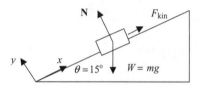

Fig. 9-7 A box sliding on a moving conveyer belt.

Therefore, the direction of the force of kinetic friction is along the positive x axis that we have chosen in the figure.

We begin by recalling the kinematic relation between velocity, acceleration, and time

$$v = v_0 + at$$

This equation will be applied to the motion of the box. To solve this problem, we need to find the time at which the velocity of box matches the velocity of the conveyer belt. The initial velocity of the box is $v_0 = 0$ m/s and so the equation we can use to find the time at which the box comes to rest on the moving conveyer belt is

$$t = \frac{v}{a} = \frac{2.44 \text{ m/s}}{a} \tag{9.24}$$

In order to find the acceleration a of the box, we will apply Newton's second law. In the direction perpendicular to the motion of the box, which we have designated the y direction, the acceleration is zero since we are making the assumption that the box stays on the conveyer belt and there is no vertical movement. Therefore, we have

$$\sum F_y = 0 = N - mg \cos \theta,$$
$$\Rightarrow N = mg \cos \theta$$

Along the direction of motion of the conveyer belt, we must take into account the force due to friction as well as due to the weight of the box. Using (9.23) Newton's second law gives us the following equation of motion

$$\sum F_x = ma = F_{\text{kin}} - mg \sin \theta = \mu_k N - mg \sin \theta$$

Therefore the acceleration can be found using

$$a = \frac{\mu_k}{m} N - g \sin \theta = \mu_k g \cos \theta - g \sin \theta$$

Putting in the numbers given in the problem, we find the acceleration is

$$a = (5/16)(9.81 \text{ m/s}^2) \cos 15° - (9.81 \text{ m/s}^2) \sin 15° = 0.42 \text{ m/s}^2$$

Therefore, the time required for the box to come to a stop, which is given by (9.24), is

$$t = \frac{2.44 \text{ m/s}}{a} = \frac{2.44 \text{ m/s}}{0.42 \text{ m/s}^2} = 5.8 \text{ s}$$

Force and Potential Energy

When a force is conservative, it can be expressed in terms of potential energy U as

$$\vec{F}(\vec{r}) = -\vec{\nabla} U \qquad (9.25)$$

In the case of a one-dimensional problem, this relation becomes

$$F(x) = -\frac{dU}{dx} \qquad (9.26)$$

EXAMPLE 9-7
A particle is moving in a region where the potential is given by

$$U(x) = x^3 - 2x^2 + \sin 3x$$

What is the force on the particle?

SOLUTION 9-7
Using (9.26), we can determine the force. Taking the derivative of the given potential, we find

$$\frac{dU}{dx} = \frac{d}{dx}(x^3 - 2x^2 + \sin 3x) = 3x^2 - 4x + 3 \cos 3x$$

And so the force is

$$F(x) = -\frac{dU}{dx} = -3x^2 + 4x - 3 \cos 3x$$

EXAMPLE 9-8
Two masses connected by a massless rope weighing 5 kg and 12 kg are resting on a plane inclined at 30°, as shown in Fig. 9-8. The coefficients of kinetic

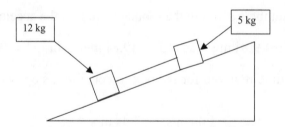

Fig. 9-8   Two masses on an inclined plane connected by a rope.

friction between each mass and the plane are 1/3 and 1/5, respectively. If the masses are released from rest find the tension T in the rope.

SOLUTION 9-8
First we consider the lower block with mass 12 kg. Let's denote the forces along the normal to the plane as perpendicular forces $F_\perp$ and the forces along the direction of motion along the plane as parallel forces $F_\parallel$. The weight of the block is

$$W = mg = (12\,\text{kg})(9.81\,\text{m/s}^2) = 117.7\,\text{N}$$

The tension T is a force that will point *up* toward the 5 kg mass. Therefore, the equations of motion for the 12 kg block are

$$\sum F_\parallel = 117.7 \sin 30° - \mu_k N_1 - T = 12a$$

$$\sum F_\perp = N_1 - 117.7 \cos 30° = 0$$

We immediately find that $N_1 = 117.7 \cos 30° = 102\,\text{N}$. Putting in 1/5 for the coefficient of kinetic friction and substituting the normal force into the equation for the parallel forces we find

$$12a = 38.5 - T \qquad\qquad (9.27)$$

where the units are in Newtons. Following a similar procedure for the second block, we obtain the following equations of motion

$$\sum F_\parallel = 49.1 \sin 30° - \mu_k N_2 + T = 5a$$

$$\sum F_\perp = N_2 - 49.1 \cos 30° = 0$$

Where we obtained $(9.81 \text{ m/s}^2)(5 \text{ kg}) = 49.1 \text{ N}$ and the tension is positive because it points from the lower block up to the 5 kg block. Solving for the normal force N_2 and setting $\mu_k = 1/3$ we find that

$$5a = 10.3 + T \tag{9.28}$$

We can solve for the tension by computing 5(9.27)–12(9.28) to get

$$0 = 68.9 - 17T$$

And we find that the tension in the rope is $T = 4.1 \text{ N}$.

The Harmonic Oscillator

We now consider a one-dimensional mass–spring system as shown in Fig. 9-9.

Imagine pulling the mass m a very small distance to the right. If this is done, a restoring force due to the spring will pull the mass back toward the left. The restoring force will be a function of displacement x. Think about your experience with springs—if you pull further to the right the force pulling the mass back to the left will be larger. In order to deduce the form that the restoring force will take mathematically, we can begin by noting it as a function of position $F = F(x)$ and expanding it in a Taylor's series

$$F(x) = F(0) + x \left(\frac{dF}{dx} \Big|_{x=0} \right) + \frac{1}{2!} x^2 \left(\frac{d^2F}{dx^2} \Big|_{x=0} \right) + \frac{1}{3!} x^3 \left(\frac{d^3F}{dx^3} \Big|_{x=0} \right)$$
$$+ \frac{1}{4!} x^4 \left(\frac{d^4F}{dx^4} \Big|_{x=0} \right) + \cdots$$

Note that a mass–spring system will have a resting equilibrium position. Since x is a small distance (think less than one), as n gets larger terms of the form x^n will go to zero. Therefore, we can throw away higher order terms and approximate

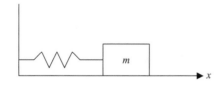

Fig. 9-9   A one-dimensional mass–spring system.

the force as

$$F(x) \approx F(0) + x \left(\frac{dF}{dx} \bigg|_{x=0} \right)$$

We can make one more approximation. Consider the leading term which is a constant. A little thought tells us that this constant must be zero. If $F(0) \neq 0$, then the mass would feel a constant force that would push it to the right if $F(0) > 0$ or constantly pull it to the left if $F(0) < 0$. This contradicts our premise that the mass has a resting equilibrium position, as you can see the mass would feel a force even without any displacement. Therefore it is clear that we must set $F(0) = 0$. Now, if we denote the first derivative of the force evaluated at $x = 0$ by

$$\left(\frac{dF}{dx} \bigg|_{x=0} \right) = -k$$

Then we obtain the familiar Hooke's law

$$F(x) = -kx \tag{9.29}$$

This is in fact the correct force law that describes the restoring force of a spring. The constant k is known as the *spring constant*. We have added the minus sign to indicate that this is a restoring force—that is, it is a force that tends to return the mass to the origin, which means it is moving in the negative x direction. Now we can use Newton's second law to write

$$F = -kx = ma$$

Or, using $a = \ddot{x} = d^2x/dt^2$ and dividing through by the mass m, we obtain

$$\frac{d^2x}{dt^2} + \frac{k}{m}x = 0 \tag{9.30}$$

This is the equation of a *simple harmonic oscillator*. The frequency of the oscillator is given by

$$\omega_0^2 = \frac{k}{m} \tag{9.31}$$

This allows us to write (9.30) in the familiar form

$$\frac{d^2x}{dt^2} + \omega_0^2 x = 0 \tag{9.32}$$

You can check in a differential equations' book that it is possible to write the solution of this equation in the form

$$x(t) = A \cos \omega_0 t + B \sin \omega_0 t \qquad (9.33)$$

To see that (9.32) is satisfied, we compute the first derivative of this expression

$$\frac{dx}{dt} = -\omega_0 A \sin \omega_0 t + \omega_0 B \cos \omega_0 t$$

So the second derivative is

$$\frac{d^2 x}{dt^2} = -\omega_0^2 A \cos \omega_0 t - \omega_0^2 B \sin \omega_0 t$$

and (9.32) is satisfied. The constants A and B are determined using the initial conditions for the position and velocity of the mass.

EXAMPLE 9-9
Consider a mass–spring system. At $t = 0$, the mass is located at $x = 1$ foot and the velocity of the mass is $v(0) = 0$. Determine the displacement and kinetic energy of the mass as a function of time.

SOLUTION 9-9
Using (9.33), we take $x(t) = A \cos \omega_0 t + B \sin \omega_0 t$. Since $\cos(0) = 1$ and $\sin(0) = 0$, at $t = 0$ we have

$$x(0) = 1 = A$$

The velocity of the mass is found by computing the time derivative of $x(t) = A \cos \omega_0 t + B \sin \omega_0 t$

$$v(t) = -\omega_0 A \sin \omega_0 t + \omega_0 B \cos \omega_0 t \qquad (9.34)$$

Therefore $v(0) = \omega_0 B$. Using the initial condition $v(0) = 0$ we conclude that $B = 0$. Therefore the displacement of the mass is

$$x(t) = \cos \omega_0 t \text{ ft}$$

The velocity is obtained from (9.34) by setting $A = 1$ and $B = 0$. This gives

$$v(t) = -\omega_0 \sin \omega_0 t$$

To find the kinetic energy, we use the basic definition

$$T = \frac{1}{2}mv^2$$

In this problem, the kinetic energy, using $v(t) = -\omega_0 \sin \omega_0 t$ turns out to be

$$T = \frac{1}{2}m\,\omega_0^2 \sin^2 \omega_0 t$$

EXAMPLE 9-10

What is the potential energy of a simple harmonic oscillator?

SOLUTION 9-10

We start with Hooke's law (9.29)

$$F(x) = -kx$$

Recalling that force is minus the gradient of the potential, or (9.26) in one dimension, we have

$$\frac{dU}{dx} = kx$$

We can integrate this equation from 0 to some arbitrary position x. This gives the form of the potential energy for the simple harmonic oscillator

$$U(x) = \frac{1}{2}kx^2 \tag{9.35}$$

EXAMPLE 9-11

What is the total energy of the mass in example 9-8?

SOLUTION 9-11

The total energy is the sum of kinetic and potential energies

$$E = T + U$$

We found that the kinetic energy was $T = \frac{1}{2}m\,\omega_0^2 \sin^2 \omega_0 t$. Using (9.35), we know that the potential energy is given by $U(x) = \frac{1}{2}kx^2$. In Example 9-8 we found that $x(t) = \cos \omega_0 t$ and so the potential energy is

$$U = \frac{1}{2}k \cos^2 \omega_0 t = \frac{1}{2}m\,\omega_0^2 \cos^2 \omega_0 t$$

The total energy is

$$E = T + U = \frac{1}{2}m\omega_0^2 \sin^2 \omega_0 t + \frac{1}{2}m\omega_0^2 \cos^2 \omega_0 t = \frac{1}{2}m\omega_0^2$$

Notice that the energy is a constant—it does not change with time and is determined by the spring constant k and the mass m.

Damping

Imagine now that the mass in our mass–spring system is in a viscous fluid that exerts a force that resists the motion. This type of force is known as a *damping force*, because it damps down the oscillations that were described in our solution (9.33). Damping is introduced into the oscillator equation via a first derivative term such that the oscillator equation is written as

$$\frac{d^2x}{dt^2} + \omega_1^2 \frac{dx}{dt} + \omega_0^2 = 0 \tag{9.36}$$

The new frequency we have introduced can be written in terms of the mass of the oscillator and a new *damping constant c* in the following way

$$\omega_1^2 = \frac{c}{m} \tag{9.37}$$

To find a solution of the differential equation, we try a solution of the form $x(t) = Ae^{st}$ where A and s are constants to be determined. Then

$$\frac{dx}{dt} = Ase^{st}$$

$$\frac{d^2x}{dt^2} = As^2 e^{st}$$

Substitution into (9.36) gives

$$As^2 e^{st} + \omega_1^2 Ase^{st} + \omega_0^2 Ae^{st} = 0$$

Dividing through by Ae^{st} gives us an equation we can use to determine s

$$s^2 + \omega_1^2 s + \omega_0^2 = 0$$

Using the quadratic formula, we obtain

$$s = \frac{-\omega_1^4 \pm \sqrt{\omega_1^4 - 4\omega_0^2}}{2} = -\omega_1^4 \pm \sqrt{\omega_1^4/4 - \omega_0^2}$$

Then the solution can be written in terms of two unknown constants A and B as

$$x(t) = Ae^{\Omega_1 t} + Be^{\Omega_2 t}$$

where

$$\Omega_1 = -\omega_1^4 + \sqrt{\omega_1^4/4 - \omega_0^2}, \quad \Omega_2 = -\omega_1^4 - \sqrt{\omega_1^4/4 - \omega_0^2}$$

The behavior of the solution is determined by the radical. First, we set it equal to zero in which case

$$\frac{\omega_1^4}{4} = \omega_0^2, \quad \Rightarrow \omega_1^2 = \frac{c_c}{m} = 2\omega_0$$

Or, we can write $c_c = 2m\omega_0$ where c_c is the critical damping coefficient. The ratio of the damping coefficient c to the critical damping coefficient is called the *damping factor*

$$d = \frac{c}{c_c}, \quad \Rightarrow \frac{c}{2m} = d\omega_0$$

The solution can be written in terms of the damping factor as

$$x(t) = A \exp\left(-d\omega_0 + \sqrt{d^2\omega_0^2 - \omega_0^2}\, t\right) + B \exp\left(-d\omega_0 - \sqrt{d^2\omega_0^2 - \omega_0^2}\, t\right)$$

$$(9.38)$$

Depending on whether the damping factor is large, small, or zero, we obtain three possible types of solution. First, we consider the case when the damping factor is large, i.e., $d \gg 1$. This case is called *overdamping*. In this situation, the radical is real and the solution is given in terms of two decaying exponentials. The motion is not periodic and the mass drifts back to its equilibrium point without any oscillations. A schematic plot of this type of solution is shown in Fig. 9-10.

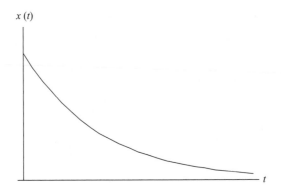

Fig. 9-10   Overdamping is just a decaying exponential.

Next, if $d \ll 1$, we have *underdamping*. In this case, the system oscillates but the oscillation decays with time. The position $x(t)$ behaves like a sinusoidal function in an envelope defined by a decaying exponential. This is because the radical is negative if $d \ll 1$ and you obtain terms of the form $e^{i\omega t}$ which can be written in terms of sine and cosine functions, multiplied by a decaying exponential of the form $e^{-d\omega_0 t}$. This is shown schematically in Fig. 9-11.

Finally, we reach the case when $d = 1$. This case is the final possibility and is called *critical damping*. In this case the solution is of the form

$$x(t) = (A + Bt)e^{-\omega_0 t}$$

The solution also decays in this case, from a peak value of $x(0) = A$. This is shown in Fig. 9-12.

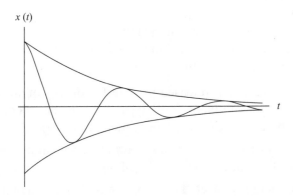

Fig. 9-11   The underdamped case oscillates, but the oscillation decays in time.

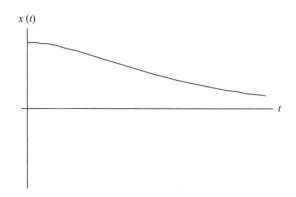

Fig. 9-12 A schematic plot of a critically damped system.

As in the overdamped case, there is no periodic or oscillatory motion. However when the system is critically damped, it returns to its equilibrium position in the fastest time possible.

Quiz

1. A bullet whose mass is $m = 7$ g is fired with a horizontal speed $v = 1000$ m/s directly at a block of wood. Ignoring friction, what is the mass of the block if its final speed is known to be 4 m/s?

2. A block weighs 11 lb. What is its mass?

3. A block weighs 11 N. What is its mass?

4. A block weighing 5 N is being pulled up a plane inclined at an angle $\theta = 35°$ with a force F_{ext} directed along the incline. Assuming that there is no friction, find
   (a) The mass of the block.
   (b) Find the normal force on the block and the minimal force that would have to be applied to keep the block from sliding downward.
   (c) The acceleration of the block when $F_{ext} = 6$ N.

5. A block of mass m is released from rest at the top of an inclined plane with coefficient of kinetic friction μ_k. What is the velocity of the block as a function of time? The plane is inclined at angle θ.

6. How long does it take the block in problem 5 to slide down to the bottom of the plane if the distance along the incline is d?

7. Two masses are connected by a rope in the same situation described in Example 9-8. This time suppose that the lower mass is $m_1 = 10$ kg and

the upper mass is $m_2 = 7$ kg with coefficients of kinetic friction 1/8 and 1/3, respectively. Write down the equations of motion and determine the tension in the rope.

8. Consider a simple harmonic oscillator. If $x(0) = 2$ m and $v(0) = 1$ m/s, what are the displacement and velocity of the mass as a function of time?

9. For the system in problem 8, what is the total energy?

10. In which type of damping does the system return to equilibrium in the quickest time?

CHAPTER 10

Circular Motion

We now consider the motion of a particle on a circular path in more detail. We begin with the simplest case, uniform circular motion and then explore angular momentum, as well as centripetal and centrifugal forces.

Uniform Circular Motion Revisited

In Chapter 8, we briefly introduced the notion of *uniform circular motion*, which we now explore in more detail. To review, if a particle is moving on a circular path with constant acceleration, we say it is undergoing uniform circular motion. As the name implies, when a particle is undergoing uniform circular motion this means that the acceleration of the particle is constant. Imagine the particle in orbit about some point P. The key feature of uniform circular motion is that there is a radial acceleration vector of the particle that points toward P at all times. We will denote the radial acceleration vector as a. Meanwhile, the velocity vector of the particle is tangential to the path. These features of the motion are illustrated in Fig. 10-1.

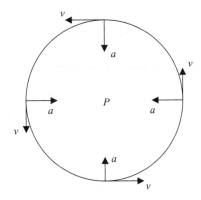

Fig. 10-1   The velocity and radial acceleration vectors for a particle in uniform circular motion about a point P.

If the particle is moving with velocity v, then the radial acceleration is related to the velocity and the radius of the orbit as

$$a = \frac{v^2}{r} \tag{10.1}$$

This type of acceleration is our first example of *centripetal acceleration.* Now let's consider circular motion with constant angular acceleration. The velocity v will not be constant but will vary with time with the *angular velocity*, which we will define in a moment.

To measure the distance and speed around the circular path, we use some basic trigonometry. Distance along the path s is given by

$$s = \theta r \tag{10.2}$$

By taking the derivative of this quantity with respect to time, we can find the angular velocity ω. The radius r of the path is constant, hence

$$\frac{ds}{dt} = \frac{d}{dt}(\theta r) = \frac{d\theta}{dt}r \tag{10.3}$$

The velocity ds/dt is just the linear velocity v. We identify the angular velocity as $\omega = d\theta/dt$, and so (10.3) tells us that

$$v = \omega r \tag{10.4}$$

This allows us to rewrite the radial acceleration defined in (10.1) in terms of angular velocity. Simple substitution of (10.4) gives

$$a_n = \omega^2 r \tag{10.5}$$

We have written this as a_n, because this is the normal component of the acceleration of the particle. Looking at (10.4), we can differentiate with respect to time to give

$$\frac{dv}{dt} = \frac{d\omega}{dt} r \tag{10.6}$$

Now $dv/dt = a$ is just the radial acceleration. This equation defines new type of acceleration called *angular acceleration* α. So we can write (10.6) as

$$a_t = \alpha r \tag{10.7}$$

This is the *tangential* component of the acceleration vector. In polar coordinates, we can denote the unit vectors as **r** and θ respectively, and so the acceleration vector is written as

$$\mathbf{a} = a_n \mathbf{N} + a_t \mathbf{T} \tag{10.8}$$

where **N** and **T** are unit vectors pointing in the normal and tangential directions of the particles path, respectively. Since the acceleration is constant, the equations for motion with constant acceleration (8.8)–(8.11) carry over in a straightforward manner for the angular variables α and ω. The relationships are the following

$$\omega = \omega_0 + \alpha t \tag{10.9}$$

$$\omega^2 = \omega_0^2 + 2\alpha\theta \tag{10.10}$$

$$\theta = \omega_0 t + \frac{1}{2}\alpha t^2 \tag{10.11}$$

$$\theta = \omega t - \frac{1}{2}\alpha t^2 \tag{10.12}$$

$$\theta = \frac{1}{2}(\omega_0 + \omega) t \tag{10.13}$$

Note that while the angular acceleration is constant in this case, the tangential and normal components of the particle's acceleration vary with time since the components depend on the angular velocity.

EXAMPLE 10-1
A flywheel has a radius of 0.45 m. The flywheel is initially at rest, then accelerates to 1800 rpm in 8 s. What is the angular acceleration?

SOLUTION 10-1
This problem can be solved using (10.9). Rearranging terms

$$\alpha = \frac{\omega - \omega_0}{t}$$

Since the flywheel starts at rest

$$\omega_0 = 0$$

The units of angular velocity as expressed in (10.3) are rad/s. So, we have to convert from revolutions per minute (rpm) to rad/s

$$\omega = \left(1800 \; \frac{\text{rev}}{\text{min}}\right)\left(\frac{1 \; \text{min}}{60 \; \text{s}}\right)\left(2\pi \; \frac{\text{rad}}{\text{rev}}\right) = 188 \; \frac{\text{rad}}{\text{s}}$$

Therefore the acceleration is

$$\alpha = \frac{\omega - \omega_0}{t} = \frac{\omega}{t} = \frac{188}{8} = 24\frac{\text{rad}}{\text{s}^2}$$

EXAMPLE 10-2
A flywheel accelerates from rest to 3200 rpm. If this takes 24 s, how many revolutions does the flywheel make in that time?

SOLUTION 10-2
Following the procedure used in Example 10-1, we start by writing the angular velocity in rad/s

$$\omega = \left(3200 \; \frac{\text{rev}}{\text{min}}\right)\left(\frac{1 \; \text{min}}{60 \; \text{s}}\right)\left(2\pi \; \frac{\text{rad}}{\text{rev}}\right) = 335 \; \frac{\text{rad}}{\text{s}}$$

Next, we need to find out how many radians are covered in this time. This can be done using (10.13). Once again, the flywheel starts from rest so $\omega_0 = 0$ and we have

$$\theta = \frac{1}{2}(\omega_0 + \omega)t = \frac{\omega t}{2} = \frac{(335 \; \text{rad/s})(24 \; \text{s})}{2} = 4020 \; \text{rad}$$

The total number of revolutions is found by dividing this quantity by 2π rad/rev

$$\theta = \frac{4020 \; \text{rad}}{2\pi \; \text{rad/rev}} = 640 \; \text{rev}$$

EXAMPLE 10-3
A flywheel accelerates from rest to 1900 rpm. What is the linear velocity of a point on the rim if the diameter of the flywheel is 1.2 m?

SOLUTION 10-3
The linear velocity is related to angular velocity using (10.4). First we write the angular velocity in rad/s

$$\omega = \left(1900 \, \frac{\text{rev}}{\text{min}}\right)\left(\frac{1}{60} \, \frac{\text{min}}{\text{s}}\right)\left(2\pi \, \frac{\text{rad}}{\text{rev}}\right) = 199 \, \frac{\text{rad}}{\text{s}}$$

The linear velocity is

$$v = \omega r = (199 \text{ rad/s})(1.2/2 \text{ m}) = 119 \text{ m/s}$$

We divided by 2 to get the radius.

EXAMPLE 10-4
A flywheel accelerates from rest to 2100 rpm in 4 s. The diameter of the flywheel is 1.3 m. What are the linear velocity and the radial acceleration?

SOLUTION 10-4
Once again, we convert the angular velocity to rad/s

$$\omega = \left(2100 \, \frac{\text{rev}}{\text{min}}\right)\left(\frac{1}{60} \, \frac{\text{min}}{\text{s}}\right)\left(2\pi \, \frac{\text{rad}}{\text{rev}}\right) = 220 \, \frac{\text{rad}}{\text{s}}$$

The linear velocity is

$$v = \omega r = (220 \text{ rad/s})(1.3/2 \text{ m}) = 143 \text{ m/s}$$

Using (10.1) the radial acceleration is

$$a = \frac{v^2}{r} = \frac{(143)^2}{(0.65)} = 31.5 \text{ km/s}^2$$

EXAMPLE 10-5
Find the linear velocity and the magnitude of the acceleration vector for the system described in Example 10-4 at $t = 0.15$ s.

SOLUTION 10-5
First we calculate the angular acceleration using (10.9). The flywheel starts from rest and reaches the stated rpm in 4 s, therefore $\omega_0 = 0$ and we find

$$\alpha = \frac{\omega - \omega_0}{t} = \frac{\omega}{t} = \frac{220}{4} = 55 \,\frac{\text{rad}}{\text{s}^2}$$

Recalling that the acceleration is constant, we can now use it to find various quantities at different times. We again apply (10.9), this time using it to solve for the velocity at an arbitrary time t. We have

$$\omega = \omega_0 + \alpha t = 0 + \left(55\,\frac{\text{rad}}{\text{s}^2} \right)(0.15\text{ s}) = 8.25\text{ rad/s}$$

The linear velocity is

$$v = \omega r = (8.25\text{ rad/s})(0.65\text{ m}) = 5.4\text{ m/s}$$

The radial acceleration from the angular acceleration can be calculated using $a = \alpha r$.

$$a_t = \alpha r = (55\text{ rad/s}^2)(0.65\text{ m}) = 36\text{ m/s}^2$$

The acceleration vector is given by (10.8). The normal component of the acceleration vector, given by (10.5) is

$$a_n = \omega^2 r = (8.25)^2(0.65) = 44\text{ m/s}^2$$

The magnitude of the acceleration vector is

$$a = \sqrt{a_n^2 + a_t^2} = \sqrt{(44\text{ m/s}^2)^2 + (36\text{ m/s}^2)^2} = 57\text{ m/s}^2$$

EXAMPLE 10-6
What angle does the acceleration vector in Example 10-5 make with the radial vector of the particle?

SOLUTION 10-6
The angle is given by

$$\phi = \tan^{-1}\left(\frac{a_t}{a_n} \right) \qquad\qquad (10.14)$$

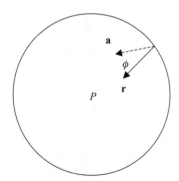

Fig. 10-2   The acceleration vector for circular motion.

In this case we have

$$\phi = \tan^{-1}\left(\frac{a_t}{a_n}\right) = \tan^{-1}\left(\frac{36}{44}\right) = 39°$$

This is illustrated in Fig. 10-2.

EXAMPLE 10-7

A wheel on a motor has a diameter of 1.4 m and an angular velocity of 1300 rpm. The motor is turned off and the wheel returns to rest in 90 s. Find the angle that the acceleration vector makes with the radial vector at 45 s.

SOLUTION 10-7

First we find the angular acceleration α, using (10.9)

$$\alpha = \frac{\omega - \omega_0}{t} = \frac{0 - (1300\,\text{rpm})\left(\frac{1}{60}\frac{\text{min}}{\text{s}}\right)\left(2\pi\frac{\text{rad}}{\text{rev}}\right)}{90\,\text{s}} = -1.5\,\text{rad/s}^2$$

The acceleration is negative because the wheel is slowing down. The tangential acceleration is

$$a_t = \alpha r = (-1.5\,\text{rad/s}^2)(0.7\,\text{m}) = -1.1\,\text{m/s}^2$$

The normal component of the acceleration is not constant, it depends on the angular velocity at a given time. Now, the initial velocity is

$$\omega_0 = (1300\,\text{rpm})\left(\frac{1}{60}\frac{\text{min}}{\text{s}}\right)\left(2\pi\frac{\text{rad}}{\text{rev}}\right) = 136\,\text{rad/s}$$

At 45 s, the angular velocity is

$$\omega = \omega_0 + \alpha t = 136\,\text{rad/s} - (1.5\,\text{rad/s}^2)(45\,\text{s}) = 68\,\text{rad/s}$$

And the normal component of the acceleration is

$$a_n = \omega^2 r = (68)^2(0.7) = 3237 \, \text{m/s}^2$$

The angle that the acceleration vector makes with the radial vector (which points to the center) is

$$\phi = \tan^{-1}\left(\frac{a_t}{a_n}\right) = \tan^{-1}\left(\frac{-1.1}{3237}\right) = -0.02°$$

Kinetic Energy of Rotation

The kinetic energy of a rigid body in rotation is defined in terms of the moment of inertia I. If the body is moving on a circular path with angular velocity ω, then the kinetic energy is

$$T = \frac{1}{2}I\omega^2 \tag{10.15}$$

EXAMPLE 10-8
Two bodies with mass m are connected by a thin rod of length L. The mass of the rod $m_r \ll m$. Find the kinetic energy of rotation if one of the bodies is fixed at the center while the other rotates about it at angular velocity ω. The situation is depicted in Fig. 10-3.

SOLUTION 10-8
Since the mass of the rod satisfies $m_r \ll m$, we can neglect it. We can find the moment of inertia using the parallel axis theorem (5.22), which we restate here

$$I_r = I_{\text{cm}} + mr^2$$

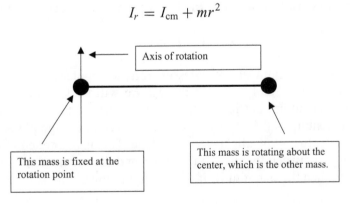

Fig. 10-3   One mass rotates about a similar fixed mass with angular velocity ω.

First we calculate the moment of inertia about the center of mass. The center of mass is at the center of the rod and therefore each particle is at a distance $L/2$. The moment of inertia is simply the discrete analog of (5.14), we replace the integral by a sum

$$I_{\text{cm}} = \sum_{i=1}^{2} m_i r_i^2 = m\left(\frac{L}{2}\right)^2 + m\left(\frac{L}{2}\right)^2 = m\frac{L^2}{2}$$

To apply the parallel axis theorem, we set r equal to the distance of the center of mass from the point of rotation. This is half the length of the rod. The mass in the parallel axis theorem is the total mass in the system, which in this case is the sum of the masses or $2m$. Using the parallel axis theorem, we find the moment of inertia to be

$$I = I_{\text{cm}} + m_{\text{tot}}r^2 = \frac{mL^2}{2} + (2m)\left(\frac{L}{2}\right)^2 = mL^2$$

Therefore the kinetic energy is

$$T = \frac{1}{2}I\omega^2 = \frac{mL^2\omega^2}{2}$$

Angular Momentum and Torque

Suppose that a mass m is in circular motion about some point P. We denote P as the origin of our coordinate system and let $\mathbf{r}$ be the radial vector from the origin to the particle. The angular momentum $\mathbf{L}$ is a vector and defined by taking the cross product of $\mathbf{r}$ with the linear momentum vector for the particle

$$\mathbf{L} = \mathbf{r} \times \mathbf{p} = \mathbf{r} \times m\mathbf{v} \tag{10.16}$$

Some jargon happy engineering professors refer to angular momentum as "moment of momentum" and denote it by $\mathbf{H_o}$. We will stick to the proper name of angular momentum, but just keep in mind these are the same thing. Since angular momentum is the cross product of two vectors in the plane of motion of the particle, it is perpendicular to that plane. This is illustrated in Fig. 10-4.

If the angle between the position vector $\mathbf{r}$ and the momentum vector of the particle is θ, then the magnitude of the angular momentum vector is

$$L = |\mathbf{L}| = rmv\sin\theta \tag{10.17}$$

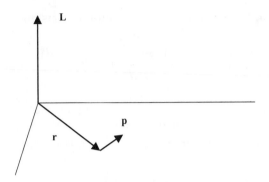

Fig. 10-4   The angular momentum vector **L** is perpendicular to the plane of motion.

In terms of components, angular momentum is

$$\mathbf{L} = \mathbf{r} \times \mathbf{p} = \begin{vmatrix} \hat{x} & \hat{y} & \hat{z} \\ x & y & z \\ p_x & p_y & p_z \end{vmatrix} = \hat{x}(yp_z - zp_y) + \hat{y}(zp_x - xp_z)$$

$$+ \hat{z}(xp_y - yp_x) \qquad (10.18)$$

Or using the fact that $p_i = dx_i/dt$

$$\mathbf{L} = \mathbf{r} \times \mathbf{p} = \hat{x}m\left(y\frac{dz}{dt} - z\frac{dy}{dt}\right) + \hat{y}m\left(z\frac{dx}{dt} - x\frac{dz}{dt}\right)$$

$$+ \hat{z}m\left(x\frac{dy}{dt} - y\frac{dx}{dt}\right) \qquad (10.19)$$

The units of angular momentum are lb-s-ft in the antiquated U.S. system and N-m-s in SI units. Suppose that the particle is being acted on by a force **F**. *Torque* is the moment of the force **F** with respect to the origin. If we denote the torque by τ, then

$$\tau = \mathbf{r} \times \mathbf{F} \qquad (10.20)$$

Newton's second law (9.5) tells us that

$$\mathbf{F} = \frac{d\mathbf{P}}{dt}$$

Therefore, we can write the torque as

$$\tau = \mathbf{r} \times \mathbf{F} = \mathbf{r} \times \frac{d\mathbf{P}}{dt}$$

Since the radius of a particle in circular motion is constant,

$$\tau = \mathbf{r} \times \frac{d\mathbf{P}}{dt} = \frac{d}{dt}(\mathbf{r} \times \mathbf{P}) = \frac{d\mathbf{L}}{dt}$$

In general, we can summarize this with an "angular Newton's law." This tells us that the sum of the torques or moments about a fixed point on a particle is equal to the time rate of change of the angular momentum

$$\sum \tau = \frac{d\mathbf{L}}{dt} \tag{10.21}$$

When no torques act on the body, (10.21) becomes

$$\frac{d\mathbf{L}}{dt} = 0$$

This tells us that if no torques act on a body, then angular momentum is conserved. It is possible to choose a coordinate system such that the torque is zero, by placing the origin of the coordinate system on the resultant of the forces acting on the body. For a system consisting of several bodies, the total angular momentum of the system is just the sum total of the angular momentum acting on each body

$$\mathbf{L}_{\text{Total}} = \sum_{i} \mathbf{L}_i \tag{10.22}$$

The *angular impulse* L_0 is the change in angular momentum from one point to another. We measure the angular momentum and velocity of the body at points 1 and 2. For a body in rotation about a fixed point

$$\Delta L = L_0 = I(\omega_2 - \omega_1) \tag{10.23}$$

where I is the moment of inertia of the body about the axis of rotation, ω_2 is the angular velocity at the second point and ω_1 is the angular velocity of the body at the first point.

EXAMPLE 10-9
A flywheel with total mass $m = 1450$ kg and radius of gyration 1.1 m is accelerated from rest to 2000 rpm in 300 s. What is the moment?

SOLUTION 10-9

In this case the angular velocities are

$$\omega_2 = \left(2000 \ \frac{\text{rev}}{\text{min}}\right)\left(\frac{1}{60} \ \frac{\text{min}}{\text{s}}\right)\left(2\pi \ \frac{\text{rad}}{\text{rev}}\right) = 209 \ \frac{\text{rad}}{\text{s}}$$

$$\omega_1 = 0 \ \frac{\text{rad}}{\text{s}}$$

We are given the radius of gyration. If we call it k, the radius of gyration is related to the moment of inertia via (5.21)

$$k = \sqrt{\frac{I}{m}}$$

Squaring both sides and rearranging terms,

$$I = mk^2$$

In this case

$$I = mk^2 = (1450 \ \text{kg})(1.1 \ \text{m})^2 = 1755 \ \text{kg-m}^2$$

Using (10.23), we have

$$\Delta L = I(\omega_2 - \omega_1) = (1755 \ \text{kg-m}^2)(209 \ \text{rad/s}) = 366{,}795 \ \text{kg-m}^2/\text{s}$$

To get the moment, we divide by the time elapsed

$$M = \frac{366{,}795 \ \text{kg-m}^2/\text{s}}{300 \ \text{s}} = 1223 \ \text{N-m}$$

EXAMPLE 10-10

A particle is in orbit about a point O. At some time t, the velocity vector of the particle is given by

$$\mathbf{v} = 2.8\mathbf{i} - 3\mathbf{j} + 2.3\mathbf{k}$$

Where velocity is measured in m/s. At this time the particle is located at $(x, y, z) = (2, -1, 3)$. If position is measured in meters, what is the magnitude of the angular momentum vector at this time? The mass of the particle is 2 kg.

SOLUTION 10-10

From the problem statement we see that SI units are being used so the units of angular momentum will be N-m-s. The position vector is

$$\mathbf{r} = 2\mathbf{i} - \mathbf{j} + 3\mathbf{k}$$

The cross product $\mathbf{r} \times \mathbf{v}$ is

$$\mathbf{r} \times \mathbf{v} = \begin{vmatrix} \mathbf{i} & \mathbf{j} & \mathbf{k} \\ 2 & -1 & 3 \\ 2.8 & -3 & 2.3 \end{vmatrix} = 6.7\mathbf{i} + 3.8\mathbf{j} - 3.2\mathbf{k}$$

To obtain the angular momentum vector, we multiply by the mass

$$\mathbf{L} = m(\mathbf{r} \times \mathbf{v}) = 2(6.7\mathbf{i} + 3.8\mathbf{j} - 3.2\mathbf{k}) = 13.4\mathbf{i} + 7.6\mathbf{j} - 6.4\mathbf{k}$$

The magnitude of this vector is

$$L = \sqrt{(13.4)^2 + (7.6)^2 + (6.4)^2} = 16.7 \text{ N-m-s}$$

EXAMPLE 10-11

For the angular momentum, position and position vectors given in Example 10-10, what is the angle between the position vector and the velocity vector?

SOLUTION 10-11

We can do this problem using (10.17). The magnitude of the position vector is

$$r = \sqrt{(2)^2 + (-1)^2 + (3)^2} = 3.7 \text{ m}$$

The magnitude of the velocity vector is

$$v = \sqrt{(2.8)^2 + (-3)^2 + (2.3)^2} = 4.7 \text{ m/s}$$

Now

$$\frac{L}{rmv} = \frac{16.7 \text{ N-m-s}}{(3.7 \text{ m})(2 \text{ kg})(4.7 \text{ m/s})} = 0.48$$

Note that this quantity is dimensionless, as it should be since it's the sine of an angle. Recall that a Newton is a kg-m^2/s to see this. The angle between the position and velocity vectors is the inverse sine of this quantity

$$\theta = \sin^{-1}(0.48) = 28.5°$$

EXAMPLE 10-12
A 3 kg mass at time t is being acted upon by a force $\mathbf{F} = -3.7\mathbf{k}$ N. Position is measured in meters and the particle is located at $\mathbf{r} = 3.5\mathbf{i} - 2\mathbf{j}$. If the velocity in m/s is given by $\mathbf{v} = 3\mathbf{i} + \mathbf{j}$, what torque is acting on the particle and what is the angular momentum about the origin?

SOLUTION 10-12
The torque is

$$\tau = \mathbf{r} \times \mathbf{F} = \begin{vmatrix} \mathbf{i} & \mathbf{j} & \mathbf{k} \\ 3.5 & -2 & 0 \\ 0 & 0 & -3.7 \end{vmatrix} = 7.4\mathbf{i} + 13.0\mathbf{j}$$

The angular momentum is

$$\mathbf{L} = m\,(\mathbf{r} \times \mathbf{v}) = 3\begin{vmatrix} \mathbf{i} & \mathbf{j} & \mathbf{k} \\ 3.5 & -2 & 0 \\ 3 & 1 & 0 \end{vmatrix} = -7.5\mathbf{k}$$

Quiz

1. A flywheel accelerates from rest to 2200 rpm in 17 s. What is the angular acceleration?

2. How many revolutions are made by the flywheel in problem 1?

3. A flywheel accelerates from rest to 2100 rpm in 4 s. The diameter of the flywheel is 1.3 m. What are the linear velocity and the radial acceleration at $t = 0.3$ s?

4. A wheel with an angular velocity of 900 rpm and a diameter of 250 mm coasts to a stop in 400 s. What is the angular deceleration? What is the linear velocity at 289 s? What is the normal acceleration at this time?

5. A particle in circular motion has tangential and normal accelerations given by $a_t = 7.8$ m/s^2 and $a_n = 18$ m/s^2. What angle does the acceleration vector make with the radial vector?

6. A rod, which initially stands upright, is pinned to the floor at the bottom end, but is free to rotate. The rod falls to the ground. If the rod is 2 m long, find its angular velocity by calculating the moment of inertia and considering the kinetic energy of the rod (*hint*: consider the potential energy of a point on the rod and use conservation of energy).

7. A flywheel with mass 2750 kg and radius of gyration 1.3 m is brought to rest from an angular speed of 175 rpm in 150 s. What moment is necessary?

8. A particle of mass 2 kg is in orbit about a point O. At some time t, the velocity vector of the particle is given by

$$\mathbf{v} = 80\mathbf{i} - 13\mathbf{j}$$

where velocity is measured in m/s. At this time the particle is located at $(x, y, z) = (3, 4, 5)$. If position is measured in meters, what is the magnitude of the angular momentum vector at this time?

9. What is the angle between the position and velocity vectors for the particle in problem 8?

10. A 5 kg mass at time t is being acted upon by a force $\mathbf{F} = 2.4\mathbf{k}$ N. Position is measured in meters and the particle is located at $\mathbf{r} = 5\mathbf{i} + \mathbf{j}$. If the velocity in m/s is given by $\mathbf{v} = 2\mathbf{i} + 2\mathbf{j}$, what torque is acting on the particle and what is the angular momentum about the origin?

CHAPTER 11

Energy and Work

Two fundamental principles of physics are that for isolated systems, energy and momentum are conserved. It turns out that if we focus on the energy in a system instead of the forces, it is often easier to solve difficult problems. We begin by considering some basic definitions and then see how to use energy to solve basic dynamics problems.

Potential and Kinetic Energy

There are two basic types of energy that we can consider with a system. The first, energy of motion, is called *kinetic energy*. When a particle or body is moving at nonrelativistic velocities, that is, velocities that are small as compared to the speed of light, the kinetic energy is related to the velocity at which the particle is moving. In this chapter, we will denote kinetic energy by T. It is related to the mass of the particle and the square of the velocity as

$$T = \frac{1}{2}mv^2 \qquad (11.1)$$

Notice that each term on the right hand side of (11.1) is positive. This tells us that kinetic energy is always positive.

The SI unit of energy is the *Joule*. Looking at (11.1), we can see that the units are

$$[T] = [m][v]^2 = \text{kg-m}^2/\text{s}^2$$
$$\Rightarrow 1 \text{ joule} = 1 \text{ kg-m}^2/\text{s}^2$$

In U.S. units, energy is measured in foot-pounds. A simple relation exists which can be used to convert energy between SI and U.S. units. This is

$$1 \text{ joule} = 0.738 \text{ ft-lb} \tag{11.2}$$

Typically, we abbreviate *joule* with a capital J, so 1 joule = 1 J. Another unit of energy, the *electron volt*, is convenient to use when dealing with atomic and subatomic systems. This is because the quantities of energy are much smaller. An electron volt is the amount of energy an electron picks up when falling through a potential of 1 volt and is given the abbreviation eV. An electron volt can be converted into joules and vice versa using

$$1 \text{ eV} = 1.6 \times 10^{-19} \text{ J} \tag{11.3}$$

EXAMPLE 11-1
A car weighing 2000 lb is traveling at 85 miles per hour. What is its kinetic energy? Give the answer in U.S. and SI units.

SOLUTION 11-1
First, we convert the velocity into feet per second

$$(85 \text{ mi/h}) (5280 \text{ ft/mi}) (1/3600 \text{ h/s}) = 125 \text{ ft/s}$$

Next we need the *mass* of the car in order to use (11.1). The weight is the mass times the gravitational acceleration, so

$$m = W/g = (2000 \text{ lb})/(32.2 \text{ ft/s}^2) = 62 \text{ slugs}$$

The kinetic energy is then

$$T = \frac{1}{2}mv^2 = \frac{1}{2}(62)(125)^2 = 484375 \text{ ft-lb}$$

We can find the kinetic energy in joules using (11.2)

$$T = 484375 \text{ ft-lb} = (484375 \text{ ft-lb}) (1 \text{ J}/0.738 \text{ ft-lb}) = 6.6 \times 10^5 \text{ J}$$

EXAMPLE 11-2
Find the kinetic energy of the planet Jupiter if the mean radius of Jupiter's orbit is 7.78×10^{11} m and the period of Jupiter's orbit is 11.94 years. The mass of Jupiter is 1.9×10^{27} kg.

SOLUTION 11-2
We convert the period of Jupiter's orbit into seconds

$$P = (11.94 \ \text{yr}) (365 \ \text{days/yr}) (24 \ \text{h/day}) (3600 \ \text{s/h}) = 3.8 \times 10^8 \ \text{s}$$

The velocity is the circumference of the orbit divided by the period

$$v = \frac{2\pi r}{P} = \frac{2\pi (7.78 \times 10^{11} \ \text{m})}{3.8 \times 10^8 \ \text{s}} = 12{,}857 \ \text{m/s}$$

Using (11.1), the kinetic energy is

$$T = \frac{1}{2}mv^2 = \frac{1}{2}(1.9 \times 10^{27} \ \text{kg}) (12{,}857 \ \text{m/s}) = 1.6 \times 10^{35} \ \text{J}$$

We'll see in a moment that there is another kind of energy called *potential energy*, but first we need to define *work*.

Forces and Work

Now let's consider the concept of *work*. By definition, work is just force multiplied by the distance over which that force is applied. That is, if we apply a force F over a distance d then the work U that is done is

$$U = Fd \qquad (11.4)$$

We see immediately from this definition that the units of work are Newton-meters if we are using the SI system, while the units of work are foot-pounds if we are using U.S. units. If the force is applied at an angle θ with respect to the line at which the object moves, then the work done is

$$U = Fd \cos \theta \qquad (11.5)$$

More formally, for a particle moving on some path, defined such that the particle is located at point P_1 at time t_1 and it's located at point P_2 at time t_2, then the work done on the particle can be calculated using

$$U = \int_C \mathbf{F} \cdot d\,\mathbf{r} \qquad (11.6)$$

The $d\mathbf{r}$ in the integrand is the infinitesimal position vector

$$d\mathbf{r} = dx\,\mathbf{i} + dy\,\mathbf{j} + dz\,\mathbf{k} \tag{11.7}$$

In Cartesian coordinates then, (11.6) can be written as

$$U = \int_C F_x dx + F_y dy + F_z dz \tag{11.8}$$

If we know the components of the particles velocity, then we can calculate the work by integration from time t_1 to t_2

$$U = \int_{t_1}^{t_2} \left(F_x \frac{dx}{dt} + F_y \frac{dy}{dt} + F_z \frac{dz}{dt} \right) dt \tag{11.9}$$

When there is rotation and the applied force constitutes a couple M, then the work done is

$$U = \int_{\theta_1}^{\theta_2} M d\theta \tag{11.10}$$

Some facts to keep in mind about work

- Notice that the units of work are the same as the units for energy.
- If the applied force acts in the same direction as the motion, then the work is positive.
- If the applied force acts in the opposite direction as the motion, then the work is negative.
- To calculate the work done by multiple forces acting on a body, find the resultant and then calculate the work as if a single force equal to the resultant were acting on the body.
- Work, like mass, is a scalar.
- A force which has no component in the direction of motion of a particle does no work.

We are now in a position to define *potential energy*. Given the symbol V, potential energy is the negative of the work done as a particle moves on a given path. That is

$$V = -\int_C \mathbf{F} \cdot d\mathbf{r} \tag{11.11}$$

In short the potential energy is the negative of the work done by the force to bring the particle to that point.

EXAMPLE 11-3
What is the potential energy of a mass m found at a height h above the earth's surface?

SOLUTION 11-3
Since potential energy is the negative of the work done, we look to (11.4) for an answer. The gravitational force is just $F_g = mg$, therefore the potential energy is just

$$V = mgh$$

EXAMPLE 11-4
A 40 lb weight is raised 20 ft. What is the potential energy of the weight at that height?

SOLUTION 11-4
Using the results of the last example, the potential energy is

$$V = mgh = \left(\frac{W}{g}\right)gh = Wh = (40 \text{ lb})(20 \text{ ft}) = 800 \text{ ft-lb}$$

This is equivalent to about a thousand joules.

EXAMPLE 11-5
A spring with spring constant $k = 11$ lb/in is compressed from 7 inches to 4 inches. How much work was done on the spring?

SOLUTION 11-5
We use the x coordinate to measure distance along the spring. First, we note that the force on a spring is given by

$$F = -kx$$

With only the x direction to worry about, (11.6) becomes

$$W = \int_{x_1}^{x_2} F(x)\,dx \tag{11.12}$$

In our case

$$W = \int_{7}^{4} (-kx)\,dx = \frac{k}{2}x^2 \bigg|_{4}^{7} = \frac{11}{2}(49 - 16) = 182 \text{ in-lb} = 15 \text{ ft-lb}.$$

EXAMPLE 11-6
A spring with spring constant $k = 18$ lb/in is compressed from 11 inches to 7 inches. What additional work is required to compress it to 4 inches?

SOLUTION 11-6
The total compression from 11 inches to 4 inches is 7 inches. The spring is then compressed an additional 4 inches. We can calculate the work by finding the work done to compress the spring 7 inches from the relaxed state and then subtracting the work required to compress the spring 4 inches from the relaxed state.

First, let's calculate the work required to compress the spring from rest by 7 inches. This is

$$U_1 = \int_0^7 (kx)\,dx = \frac{k}{2}x^2 \Big|_0^7 = \frac{18}{2}(49) = 441 \text{ in-lb} = 37 \text{ ft-lb}$$

Now we calculate the work required to compress the spring from rest by 4 inches, which is

$$U_2 = \int_0^4 (kx)\,dx = \frac{k}{2}x^2 \Big|_0^7 = \frac{18}{2}(16) = 144 \text{ in-lb} = 12 \text{ ft-lb}$$

The work done compressing the spring from 11 inches to 7 inches and then from 7 inches to 4 inches is the difference of these quantities

$$U = U_2 - U_1 = 37 \text{ ft-lb} - 12 \text{ ft-lb} = 25 \text{ ft-lb}$$

The Work–Energy Theorem

The work–energy theorem tells us that the work done on a particle in motion is given by the change in the particle's kinetic energy. This can be written simply as

$$U = T_f - T_i = \Delta T \tag{11.13}$$

In words, we say that the change in the kinetic energy of a particle is equal to the work done by all forces acting on that particle.

EXAMPLE 11-7
A ladder with mass m and length l is standing vertically. If falls to the floor. What is the angular velocity of the ladder when it hits the floor? Suppose that the ladder is 10 ft long and weighs 50 lb.

SOLUTION 11-7

From the work–energy theorem, we know that the work is equal to the change in kinetic energy. Initially, the ladder is at rest therefore its velocity is zero. So (11.13) gives us

$$U = T_f$$

The moment of inertia can be taken to be that of a rod, so we have

$$I = \frac{1}{3}ml^2$$

And we take the kinetic energy to be $T = \frac{1}{2}I\omega^2$ where ω is the angular velocity of the ladder. Now, what is the work done? It's force times distance. The work is done by the gravitational force, and this is just mass times the gravitational acceleration times the height covered. The distance we are interested in is the distance that the center of mass travels. It travels from the midpoint of the ladder standing upright which is $h = \frac{1}{2}l$ to the floor where we take $h = 0$. So the work done is

$$U = \frac{1}{2}mgl$$

Setting this equal to the final kinetic energy by the work–energy theorem, we have

$$\frac{1}{2}mgl = \frac{1}{6}ml^2\omega^2$$

Solving, we find an expression for the angular velocity

$$\omega = \sqrt{\frac{3g}{l}}$$

EXAMPLE 11-8

At the gym a trainer lifts a 50 lb dumbbell 4 ft to place it in a storage bin. How much work does the trainer do on the dumbbell? How much work does the force of gravity do on the dumbbell? Assume that the trainer lifts the weight carefully enough so that it does not accelerate.

SOLUTION 11-8

Since the weight isn't accelerated, its in equilibrium and the sum of the forces acting on the weight are zero. The forces acting on the weight are the force of gravity, which points down and the force exerted by the trainer. Since the sum of the forces is zero, the force exerted by the trainer is equal in magnitude but opposite in direction to the force of gravity, which is

$$F = mg = 50 \text{ lb}$$

The trainer is exerting an upward force to lift the weight. The displacement vector points along the upward direction, so the angle between the applied force and the displacement vector is $\theta = 0$. That means we can use (11.4) to calculate the work done by the trainer. It is

$$U = Fd = (50 \text{ lb})(4 \text{ ft}) = 200 \text{ ft-lb}$$

The force of gravity acts in the opposite direction to the displacement vector of the weight. Looking at (11.5), we see that $\theta = 180°$ is the appropriate choice, giving

$$U = Fd \cos \theta = (50 \text{ lb})(4 \text{ ft}) \cos 180° = -200 \text{ ft-lb}$$

Not surprisingly, the work done by gravity is opposite in sign to the work done by the trainer in lifting the weight.

EXAMPLE 11-9

A worker is pulling a crate with a mass $m = 25$ kg across a frictionless surface. He is using a rope over his shoulder which is at an angle $\theta = 25°$ with respect to the horizontal. What is the minimum force he must apply to keep the crate moving? How much work is required to pull the crate a distance $d = 17$ m?

SOLUTION 11-9

The normal force on the crate is $N = mg = (25)(9.81) = 245$ N. The component along the direction of motion is

$$F = N \cos 25° = (245) \cos 25° = 222 \text{ N}$$

This can be understood by looking at Fig. 11-1. The horizontal component of the normal force, F_h, which is along the direction of the displacement vector is the force that must be overcome in order for motion to occur. This is found using trigonometry. Since the worker is surprisingly applying the minimum force required to move the crate, we take this as the force applied on the rope.

Now that we know the applied force, this problem can be solved using (11.5). We have

$$U = Fd \cos \theta = (222 \text{ N})(17 \text{ m}) \cos 25° = 3420 \text{ J} = 3.4 \text{ kJ}$$

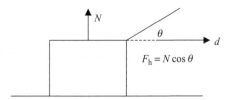

Fig. 11-1   The horizontal component of the force applied to the rope is what does the work, since it is along the direction of the displacement vector.

Energy and Momentum

Looking at the formula for kinetic energy (11.1) it should be apparent that we can easily rewrite the expression for kinetic energy in terms of momentum. Recall that momentum at nonrelativistic velocities is written as $\mathbf{p} = m\mathbf{v}$ or just $p = mv$ for motion in one dimension. For simplicity we consider the one-dimensional case. Then

$$T = \frac{1}{2}mv^2 = \frac{1}{2}\left(\frac{m}{m}\right)mv^2 = \frac{1}{2m}(mv)^2 = \frac{p^2}{2m} \tag{11.14}$$

EXAMPLE 11-10
The mass of the neutron is 1.67×10^{-27} kg. A neutron is moving through a laboratory and its energy is found to be 6.1×10^{-15} joules. Find the momentum and velocity of the neutron.

SOLUTION 11-10
Solving for momentum using (11.14), we have

$$p = \pm\sqrt{2mT}$$

There is not enough information given in the problem to determine the direction of the momentum, so we will just solve for the magnitude of the momentum and ignore the sign. We find

$$p = \sqrt{2mT} = \sqrt{2\,(1.67 \times 10^{-27})\,(6.1 \times 10^{-15})} = 4.5 \times 10^{-21}\ \text{kg}\,\frac{\text{m}}{\text{s}}$$

The velocity of the neutron is

$$v = \frac{p}{m} = \frac{4.5 \times 10^{-21}}{1.67 \times 10^{-27}} = 2.7 \times 10^{6}\,\frac{\text{m}}{\text{s}}$$

Power

The rate of doing work is known as *power*. Mathematically speaking, power is the rate of change of work with respect to time. If we denote power by P then it is defined as

$$P = \frac{dU}{dt} \tag{11.15}$$

In terms of force, power is defined as

$$P = \mathbf{F} \cdot \frac{d\mathbf{r}}{dt} = \mathbf{F} \cdot \mathbf{v} \tag{11.16}$$

Since it is defined as the dot product of two vectors, power is a scalar. The SI unit of power is the joule per second which is called a *Watt*

$$1\,\frac{J}{s} = 1\ \text{Watt} \tag{11.17}$$

Power can also be described using U.S. units. In this case, the unit of power is the foot pound per second (ft-lb/s). It is more common to see power measured in *horsepower*, where

$$1\ \text{hp} = 550\ \text{ft-lb/s} \tag{11.18}$$

For a couple, power is expressed as

$$P = \mathbf{M} \cdot \omega \tag{11.19}$$

where $\mathbf{M}$ is the moment of the couple.

EXAMPLE 11-11
A car that weighs 2500 lb is traveling at 85 miles per hour. The coefficient of kinetic friction between the car and the road is 0.6. What is the kinetic energy of the car? What horsepower is the car developing?

SOLUTION 11-11
First, we convert the speed of the car into feet per second

$$\left(85\frac{\text{mi}}{\text{h}}\right)\left(\frac{5280\ \text{ft}}{\text{mi}}\right)\left(\frac{1\ \text{h}}{3600\ \text{s}}\right) = 127\frac{\text{ft}}{\text{s}}$$

To find the kinetic energy, we need the mass, which is

$$m = \frac{W}{g} = \frac{2400\ \text{lb}}{32.2\ \text{ft/s}^2} = 78\ \text{slugs}$$

The kinetic energy is

$$T = \frac{1}{2}mv^2 = \frac{1}{2}(78)(127)^2 = 6.1 \times 10^5 \text{ ft-lb}$$

The only force acting in the direction of the cars motion is the frictional force, which is

$$f = \mu_k N = \mu_k W = (0.6)(2500) = 1500 \text{ lb}$$

The force generated by the car is equal to this value. While the frictional force points in the opposite direction of the car's motion and is therefore at an angle $\theta = 180°$ with the velocity vector, the force generated by the car is parallel to the velocity vector. Therefore the angle between the force generated by the car is $\theta = 0°$. Using (11.16), we find the power to be

$$P = |F||v|\cos\theta = (1500)(127) = 1.9 \times 10^5 \frac{\text{ft-lb}}{\text{s}}$$

In horsepower, this is

$$P = \left(1.9 \times 10^5 \frac{\text{ft-lb}}{\text{s}}\right) \bigg/ (550 \text{ ft-lb/hp}) = 346 \text{ hp}$$

EXAMPLE 11-12
An experimental rocket car is pulled down a track with a force of 67 kN. The speed of the car is 247 km/h. How much power is being developed?

SOLUTION 11-12
First, we convert the speed into meters per second

$$\left(247 \frac{\text{km}}{\text{h}}\right)\left(1000 \frac{\text{m}}{\text{s}}\right)\left(\frac{1 \text{ h}}{3600 \text{ s}}\right) = 68.6 \frac{\text{m}}{\text{s}}$$

When the force acts along the direction of motion, power is just force times velocity

$$P = Fv = (67,000 \text{ N})\left(68.6 \frac{\text{m}}{\text{s}}\right) = 4.6 \text{ MW}$$

Efficiency

Efficiency is defined as work output divided by work input or power output divided by power input.

$$E = \frac{P_{\text{out}}}{P_{\text{in}}} \tag{11.20}$$

The power output is always less than the input power, therefore the efficiency is a fraction less than one. It is often expressed as a percentage.

The efficiency of an engine is often given as a percentage. If the efficiency at which an engine operates is given, calculate power in the normal way and then divide by the efficiency to find the power of the engine.

EXAMPLE 11-13
A robot invented by a foreign automobile company raises a 2500 kg mass straight up. The velocity is 7 m/s. If the efficiency of the robot's engine is known to be 78 percent, what is the power of the engine?

SOLUTION 11-13
The force acting on the mass is the force of gravity, which is

$$F = (2500 \text{ kg})(9.81 \text{m/s}^2) = 24.5 \text{ kN}$$

Without taking efficiency into account, the power is

$$P = Fv = (24,500 \text{ N})(7 \text{ m/s}) = 171.5 \text{ kW}$$

To find the real power of the engine, we divide this by the efficiency

$$P = \frac{171.5}{0.78} = 219 \text{ kW}$$

EXAMPLE 11-14
The power indicated for a certain engine is 500 kW. A test measures the power output as 423 kW. What is the efficiency of the engine?

SOLUTION 11-14

$$E = \frac{P_{\text{out}}}{P_{\text{in}}} = \frac{423}{500} = 0.85 \text{ or } 85 \text{ percent}$$

Basic Conservation Laws

Conservation laws are at the heart of physics and stand as fundamental "truths" as to how the physical world operates. Two key conservation laws that are important in dynamics are *conservation of energy* and *conservation of momentum*.

In order to apply a conservation law to a system, that system has to be *isolated*. What we mean by this is that the system is completely free of outside influences. No system can be *truly* isolated, however we can consider a system isolated for practical purposes. For example, you might want to study the temperature of some process. You can consider it isolated by thermally insulating it from the outside world so that little heat energy enters or leaves the system. A criterion you could use to determine whether the system is isolated is the following. Is the insulation good enough so that the heat entering the system from the outside world is significantly smaller than the temperatures you are trying to measure? If the answer is yes then the system can be considered isolated and you can apply conservation laws to analyze the system.

The key aspect of conservation laws is that some measured quantity M remains constant. That is if

$$M = \text{constant}$$

then we say that M is *conserved*. The quantity of interest remains constant for the *overall* system, but it may redistribute itself in different ways *throughout* the system. In a mechanical system, kinetic energy might be converted into heat. If the system is well insulated and the heat is not allowed to escape, the total energy in the system would be the same but it has been redistributed, with some of the kinetic energy now heat energy.

A quintessential example often used is the redistribution of kinetic and potential energy in a mass spring system. What we are about to describe is not a realistic system, it is a thought experiment to get the concept of energy conservation across. We imagine that the system is totally isolated and is ideal so that energy is not dissipated eventually bringing the spring to rest.

Imagine that the total energy in the system is E. The total energy in a system is the sum of kinetic and potential energies

$$E = T + V \tag{11.21}$$

We take the x coordinate such that x vanishes when the spring is completely at rest. This is shown in Fig. 11-2.

At different points x in the motion of the mass, energy will be distributed among kinetic and potential energy in different ways. For example, the mass will

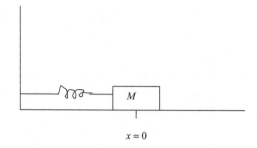

Fig. 11-2 The mass is at rest and centered about $x = 0$. This is when the spring is uncompressed and not extended.

reach turning points when the spring is maximally compressed and maximally extended. At those instants, the velocity of the mass v will be zero, therefore the kinetic energy is

$$T = \frac{1}{2}mv^2 = 0$$

At that instant, all of the energy of the system will be potential energy. We saw in Chapter 9 that the potential energy of the spring is

$$V = \frac{kx^2}{2}$$

Assuming that the mass is oscillating with some energy E, when it passes the point $x = 0$, the point at which the spring is in its natural configuration, we can see from $V = \frac{kx^2}{2}$ that the potential energy is zero. Therefore all of the energy is *kinetic*. Since the system has total energy E, the speed is

$$v = \sqrt{\frac{2E}{m}}.$$

The sign will be positive if the mass is moving in the direction in which the spring is extended, and will be negative if the mass is moving in the direction at which the spring is compressed at that instant.

When the spring is maximally compressed or extended, then at that instant the velocity is zero, as we described above. To find the length of the spring at that time, we set the potential energy equal to the total energy E. For a given energy E, the spring could be extended to a length

$$l = \sqrt{\frac{2E}{k}}$$

At all times, E is a fixed constant value if energy is conserved. We can use the relation $E = T + V$ to find out how the energy is distributed among kinetic and potential energy.

EXAMPLE 11-15
A ball rolls down a hill that is 70 m above level ground, and rolls up onto a shorter hill. There is no friction between the ball and the ground. When the ball reaches the smaller hill, its velocity is 7 m/s. What is the height of the smaller hill?

SOLUTION 11-15
We can apply conservation of energy in this case in the following way. If we assume that the energy is constant, which we can do because there is no friction, then the sum of the change in kinetic energy and the change in potential energy is zero, that is since at some point

$$E = \frac{1}{2}mv_1^2 + mgy_1$$

While at another point in time

$$E = \frac{1}{2}mv_2^2 + mgy_2$$

The change $\Delta E = 0$ and we can write

$$0 = \Delta T + \Delta V \qquad (11.22)$$

where T and V are the kinetic and potential energies, respectively.
    At the top of the first hill, the velocity of the ball is zero and the energy of the ball is all potential energy. The potential energy of a mass m in a gravitational field at a height y is

$$V = -mgy \qquad (11.23)$$

The change in potential energy is the final potential energy, which is at the unknown height h of the smaller hill, minus the potential energy at the initial point

$$\Delta V = -mgh - [-mg(70)] = mg(70 - h)$$

The initial kinetic energy is zero since the velocity of the ball is zero at the start, while it's $\frac{1}{2}mv^2$ at the top of the smaller hill. We apply conservation of energy and equate kinetic and potential energies

$$\frac{1}{2}mv^2 = mg\,(70 - h)$$

Canceling the mass we obtain

$$\frac{1}{2}v^2 = g\,(70 - h)$$

Solving for h gives

$$h = \frac{70\,g - \dfrac{1}{2}v^2}{70} = 9.5 \text{ m}$$

When a system is isolated, the total momentum in the system is also conserved. This means that

$$p = \text{constant} \qquad\qquad (11.24)$$

EXAMPLE 11-16
A bullet with a mass $m = 7.2$ g is fired with a speed $v = 1500$ m/s into a wood block with a mass $M = 13$ kg resting on a frictionless surface. What is the velocity of the block after the bullet becomes lodged in it?

SOLUTION 11-16
Let the subscript b denote the bullet and the subscript B denote the block. Initially the total momentum is

$$p_i = mv_b + Mv_B$$

The velocity of the block is zero initially, so $p_i = m_b v_b$. At the final point, the block and the bullet form one system, so the final momentum is

$$p_f = (m + M)v$$

Conservation of momentum requires that $p_f = p_i$ so we have

$$mv_b = (m + M)v, \Rightarrow$$

$$v = \frac{mv_b}{(m + M)} = \frac{10.8 \text{ kg } \frac{\text{m}}{\text{s}}}{\left(\dfrac{7.2}{1000}\text{kg} + 13 \text{ kg}\right)} = 0.8 \text{ m/s}$$

Quiz

1. Find the kinetic energy of the planet Mars if the mean radius of its orbit is 2.3×10^{11} m and the period of Mars' orbit is 1.88 years. The mass of Mars is 6.4×10^{23} kg.

2. A man is riding a horse at 30 miles per hour. The man weighs 170 lb and the horse weighs 1500 lb. What is their kinetic energy in Joules?

3. A 50 lb weight is raised 35 ft. What is the potential energy of the weight at that height?

4. What is the work done compressing a spring from its relaxed state by 6 inches if the spring constant is 20 lb/in?

5. A worker is pulling a crate with a mass $m = 50$ kg across a frictionless surface. He is using a rope over his shoulder which is at an angle $\theta = 35°$ with respect to the horizontal. What is the minimum force he must apply to keep the crate moving? How much work is required to pull the crate a distance $d = 20$ m?

6. The kinetic energy of an electron in an aluminum wire is 5×10^{-18} J. What is the electron's velocity? The mass of an electron is 9.11×10^{-31} kg.

7. The kinetic energy of a proton is 5×10^{-18} J. What is the momentum of the proton? What is its velocity? The mass of the proton is 1.67×10^{-27} kg.

8. A train car is pulled down a track with a force of 37 kN. The speed of the car is 17 km/h. How much power is being developed?

9. The power indicated for a certain engine is 300 kW. A test measures the power output as 260 kW. What is the efficiency of the engine?

10. A secret lab in Geneva has a test slide used to accelerate weapons. The slide is 45 m above level ground. A ball, initially at rest, rolls down without friction. What is the speed of the ball when it reaches the bottom of the slide?

CHAPTER 12

Waves and Vibrations

In this chapter, we will study the mechanical behavior of waves and vibrations. We begin with some general remarks on mechanical waves, focusing on the basic but fundamental type of wave described by a sinusoidal oscillation. Then we consider some properties of vibrating mechanical systems.

Basic Wave Motion

A *wave* is a disturbance that propagates through space. Many waves, such as water waves, require a mechanical medium to propagate through. Another example, one of the most familiar examples of a mechanical wave in daily life, is sound waves which require air for propagation. Not all waves require a material medium to propagate through, however. For example, electromagnetic radiation including light and radio waves can propagate through a vacuum. Electromagnetic waves are not a mechanical disturbance of any kind, although they share many of the general properties of waves that we will discuss in this chapter.

A great deal can be learned about waves by focusing on a simple vibrating string. We imagine that at one end of the string a person shakes the string up

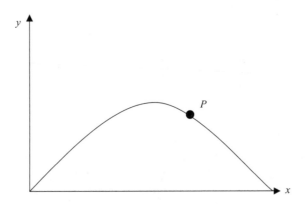

Fig. 12-1   Following the motion of a point on a vibrating string.

and down, causing a wave to propagate across it. We pick a point P at some point along the string at a given time during its motion, as shown in Fig. 12-1.

The position or height of the point P is a function of the position x and time t as the string moves up and down. Functionally, we can write

$$y = y(x, t) \tag{12.1}$$

There are two quantities of interest that we can use to describe the string. The first is the slope at any given time. The slope is given by the partial derivative of y with respect to x

$$\text{slope} = \frac{\partial y}{\partial x} \tag{12.2}$$

The velocity of the height of the point is

$$v_P = \frac{\partial y}{\partial t} \tag{12.3}$$

Generally speaking, any function that can be written in the form

$$f(x, t) = f(x - vt) \tag{12.4}$$

modulo some constants multiplying the argument describes a traveling wave. More specifically, a wave with argument of the form (12.4) is one that is traveling to the *right*, or towards more positive x. A wave that is traveling in the opposite direction, toward the left or towards more negative x, is of the form

$$g(x, t) = g(x + vt) \tag{12.5}$$

These ideas are illustrated in Fig. 12-2 and Fig. 12-3.

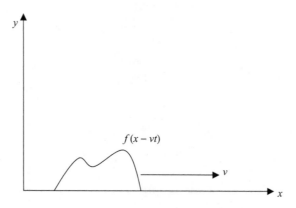

Fig. 12-2 A wave traveling to the right has the functional form $f(x - vt)$. As time increases, the wave travels to the right.

Sinusoidal Waves

A waveform you are probably already familiar with is the *sinusoidal* wave. This is a basic waveform that arises frequently in the study of engineering mechanics, electromagnetics, and physics. A sin wave is shown in Fig. 12-4.

The functional form of a sin wave traveling to the right is, using (12.4) as a guide

$$f(x,t) = A \sin(kx - \omega t) \tag{12.6}$$

The argument of the sin function, $\phi = kx - \omega t$, is called the *phase* of the wave. The phase is constant, which will be important later.

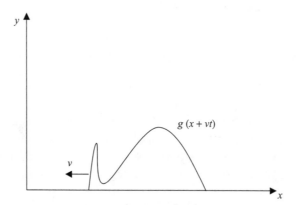

Fig. 12-3 A wave traveling to the left has the functional form $g(x,t) = g(x + vt)$. As time increases, the wave travels to the left.

Sinx

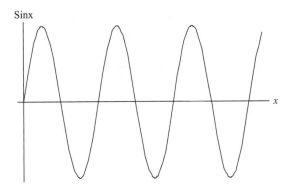

Fig. 12-4   A sin wave.

The multiplicative constant A is called the amplitude of the wave. This is the maximum value that the wave can attain since

$$-1 \leq \sin \theta \leq 1$$

The amplitude has dimension of length, in SI units we can measure the amplitude in meters, although different units may be convenient depending on the type of wave. The amplitude of the wave is basically the maximum height of the wave above the origin. This is shown in Fig. 12-5.

Returning to the phase, let's determine the units of the multiplicative constants. The phase is

$$\phi = kx - \omega t$$

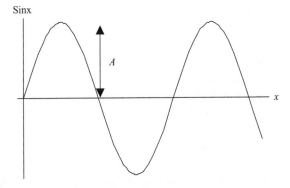

Fig. 12-5   The amplitude of a sin wave is the max height above the origin.

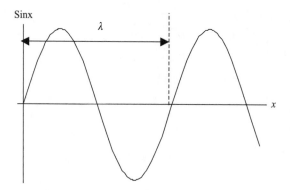

Fig. 12-6 The wavelength is the shortest distance covered by the pattern of the wave, which repeats itself at regular intervals.

Since the phase is just the argument of the sin function, it is measured in radians. So, we need to cancel the dimensions of length and time seen in the functional form of the phase. The dimensions of k are

$$[k] = \frac{\text{rad}}{[x]} = \frac{\text{rad}}{\text{length}}$$

If we are using SI units, then the units of k will be radians per meter. The constant k is called the *wave number*. It is related to another quantity denoted by the mysterious Greek symbol λ that scientists call the *wavelength*. The relationship between the two is given by

$$k = \frac{2\pi}{\lambda} \tag{12.7}$$

In SI units, wavelength is measured in meters. The wavelength λ is the shortest distance over which the form or pattern of the wave repeats itself. This is illustrated in Fig. 12-6.

The *period T* is the time interval required for one pattern of the wave to pass a fixed point. Or another way to say this is that it is the time interval required for the wave to repeat itself. Mathematically, this is the time required to pass such that the functional form of the wave assumes the same value

$$f(x,t) = f(x,t+T) \tag{12.8}$$

The SI unit for period is the second. Inverting the period gives us a familiar measure of the properties of the wave, the *frequency*. First, we define the *angular*

frequency as

$$\omega = \frac{2\pi}{T} \text{ rad/s} \tag{12.9}$$

A more familiar form is the plain old *frequency*. Denoted by another mysterious Greek symbol ν, frequency is the number of cycles made by the wave per unit time at a given point. Frequency is just the inverse of the period

$$\nu = \frac{1}{T} \tag{12.10}$$

The units of frequency are cycles per second. Formally, this is called the *hertz*

$$1 \text{ hertz} = 1 \text{ cycle/s} \tag{12.11}$$

As can be seen by comparing (12.9) and (12.10), frequency (plain Jane) is related to angular frequency via

$$\nu = \frac{\omega}{2\pi} \tag{12.12}$$

The propagation of a wave is governed by the *wave equation*. The wave equation is a partial differential equation relating spatial and temporal variations of the wave. In one dimension, if the wave is propagating along the x direction, the wave equation is

$$\frac{\partial^2 f}{\partial x^2} = \frac{1}{v^2} \frac{\partial^2 f}{\partial t^2} \tag{12.13}$$

How fast is the wave moving? We can find the *phase velocity* of the wave by setting the phase equal to some constant C and then differentiating. First, let's remind ourselves that the phase is constant

$$\phi = kx - \omega t = C$$

Taking the time derivative, we have

$$k\frac{dx}{dt} - \omega = 0$$

Solving for the velocity $v = dx/dt$ we find

$$v = \frac{\omega}{k} \tag{12.14}$$

Now $\omega > 0$ and $k > 0$, therefore the velocity $dx/dt > 0$. That tells us that the wave is traveling to the right, or towards more positive x. This is consistent with what we stated earlier about a wave with argument $x - vt$.

Dispersion

Unfortunately things don't always stay simple in the physical world. One example is the phenomenon of dispersion. Sometimes waves with different frequencies travel at different speeds. A good example is an electromagnetic wave passing through some kind of medium. This basically means that a group of waves that was initially traveling together *disperses*, with each frequency component traveling with a different speed. The most common example of this is white light passing through a prism that splits the light out into the different primary colors.

In such a case the passing wave can be described by a wave packet, which is a *packet* with an overall shape which is filled with individual waves made up of the different frequency components (see Fig. 12-7). Each of these is moving at a different phase velocity ω/k.

The packet, meanwhile, has a relationship between angular frequency and wave number that is a functional form $\omega = \omega(k)$ called a *dispersion relation*. The motion of the packet as a whole is described the group velocity, which is

$$v_g = \frac{d\omega}{dk} \tag{12.15}$$

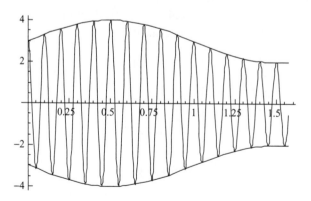

Fig. 12-7 A simple example of a wave packet.

When you hear about experiments with light where seemingly anomalous behavior occurs, such as components of the wave traveling "faster than the speed of light," you are hearing an example of differences between group and phase velocity. In these cases, the phase velocity can exceed the speed of light, but it makes no difference because the waves only travel at the phase velocity inside the wave packet. For you, the observer, you can only detect the leading edge of the wave packet which is moving at the group velocity. It turns out that the group velocity travels at the speed of light.

EXAMPLE 12-1
A wave is described by $f(x,t) = 0.02 \sin(150x - 3.8t)$. What are the amplitude, wavelength, and period of the wave? What is the frequency? What is the speed of the wave? Which direction is the wave traveling?

SOLUTION 12-1
Comparison with (12.6) tells us that the amplitude of the wave is $A = 0.02$ m $= 2$ cm. The wave number is $k = 150$ and the angular frequency is $\omega = 3.8$ rad/s. Using (12.7), we find the wavelength to be

$$\lambda = \frac{2\pi}{150} = 0.042 \text{ m} = 4.2 \text{ cm}$$

We can find the period from the angular frequency using (12.9). Rewriting this formula gives

$$T = \frac{2\pi}{\omega} = \frac{2\pi}{3.8} = 1.7 \text{ s}$$

The frequency is

$$\nu = \frac{1}{T} = \frac{1}{1.7 \text{ s}} = 0.61 \text{ Hz}$$

The velocity of the wave can be calculated from (12.14), from which we find

$$v = \frac{\omega}{k} = \frac{3.8}{150} = 0.03 \text{ m/s} = 30 \text{ cm/s}$$

EXAMPLE 12-2
In quantum mechanics, the energy and momentum of a free particle of mass m, which is described by a wave packet, are related to the angular frequency and wave number via the relations $E = \hbar\omega$ and $p = \hbar k$. Show that the phase velocity of the wave packet does not give the correct velocity of the particle but that the group velocity does.

SOLUTION 12-2

We assume nonrelativistic motion for which the kinetic energy of the particle is related to the momentum using

$$E = \frac{p^2}{2m}$$

Substitution of $E = \hbar\omega$ and $p = \hbar k$ allows us to write

$$\hbar\omega = \frac{\hbar^2 k^2}{2m}$$

Solving for ω/k

$$\frac{\omega}{k} = v_{phase} = \frac{\hbar k}{2m} = \frac{p}{2m}$$

Using $p = mv$, we find the relationship of the phase velocity to the velocity of the particle to be

$$v_{phase} = \frac{p}{2m} = \frac{mv}{2m} = \frac{v}{2}$$

For a quantum mechanical wave packet then, the phase velocity is exactly one half the velocity of the particle. What about the group velocity? Returning to

$$\hbar\omega = \frac{\hbar^2 k^2}{2m}$$

We solve for the angular frequency

$$\omega = \frac{\hbar k^2}{2m}$$

Now we compute the derivative

$$v_g = \frac{d\omega}{dk} = \frac{d}{dk}\left(\frac{\hbar k^2}{2m}\right) = \frac{\hbar k}{m} = \frac{p}{m} = \frac{mv}{m} = v$$

Therefore we see that for a free quantum particle, which is described by a wave packet, the group velocity is equal to the classical velocity of the particle.

For waves on a string, the velocity of the wave is related to the mass density σ and tension T in the string as

$$v = \sqrt{\frac{T}{\sigma}} \tag{12.16}$$

The mass density is a *linear* mass density, and so the units are mass per unit length.

EXAMPLE 12-3
A climber who weighs 175 lb and his friend are at a cliff that is 34 m above the valley below. The climber is hanging off the cliff with a rope that is 32 m in length, while his friend waits at the top of the cliff. The friend, thinking he is funny, shakes the rope vigorously. If the mass density of the rope is 820 g/m, how long does it take for the disturbance to travel down to the climber, causing him to lose his grip, falling to the valley below and breaking his ankle?

SOLUTION 12-3
First, let's convert the weight of the climber into SI units. Since 1 lb is about 4.45 N, the climber's weight is

$$(175 \text{ lb})(4.45 \text{ N/lb}) = 779 \text{ N}$$

This corresponds to a mass of about 79.4 kg. Is the body mass index of the climber too high? There is not enough information in the problem to tell. Nonetheless, the tension is just equal to the weight of the climber, and so

$$\frac{T}{\sigma} = \frac{779 \text{ N}}{0.82 \text{ kg/m}} = 950 \frac{\text{m}^2}{\text{s}^2}$$

The velocity of the disturbance traveling through the rope is

$$v = \sqrt{\frac{T}{\sigma}} = \sqrt{950} = 30.8 \text{ m/s}$$

Hence, the time required for the disturbance to reach the climber is

$$t = \frac{32 \text{ m}}{30.8 \text{ m/s}} = 1.04 \text{ s}$$

Superposition and Interference

Waves have a special property. When two waves pass each other or collide, they add up to form a new wave. This ability to add two waves together and obtain a new third wave is called *superposition*. As an example, imagine two water waves that meet and then form a new wave that may have different characteristics than the waves that went together to form it. Sometimes if things are just right the waves completely cancel and there is no wave that results.

For sinusoidal waves, it is easy to characterize the properties of the new wave that is formed by the superposition. For simplicity we will stick to waves that have the same amplitude and consider adding together two waves. Recalling the trig identity

$$\sin A + \sin B = 2 \sin \left(\frac{A + B}{2} \right) \cos \left(\frac{A - B}{2} \right) \tag{12.17}$$

If we let $f_1(x, t) = A \sin(k_1 x - \omega_1 t)$ and $f_2(x, t) = A \sin(k_2 x - \omega_2 t)$ then the sum or superposition $f_3 = f_1 + f_2$ is described by

$$f_3(x, t) = 2A \sin \left[\frac{(k_1 + k_2)}{2} x - \frac{(\omega_1 + \omega_2)}{2} t \right]$$

$$\times \cos \left[\frac{(k_1 - k_2)}{2} x - \frac{(\omega_1 - \omega_2)}{2} t \right] \tag{12.18}$$

Notice that the amplitude of the combined wave is exactly twice that of the amplitudes of the original waves—but we considered the special case of adding two waves that had the same amplitudes.

EXAMPLE 12-4
Two waves $f_1(x, t) = 0.04 \sin(70x - 3t)$ and $f_2(x, t) = 0.04 \sin(67x - 5t)$ interfere. Describe the resulting wave.

SOLUTION 12-4
The amplitude of the wave that results from superposition is

$$A = 2(0.04) \text{ m} = 0.08 \text{ m} = 8 \text{ cm}$$

We have

$$\frac{k_1 + k_2}{2} = \frac{70 + 67}{2} = 68.5$$

$$\frac{k_1 - k_2}{2} = \frac{70 - 67}{2} = 1.5$$

And

$$\frac{\omega_1 + \omega_2}{2} = \frac{3 + 5}{2} = 4$$

$$\frac{\omega_1 - \omega_2}{2} = \frac{3 - 5}{2} = -1$$

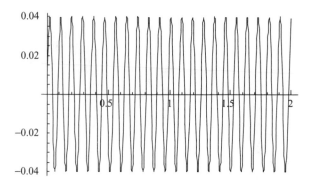

Fig. 12-8 A plot of $f_1(x,t) = 0.04 \sin(70x - 3t)$ at time $t = 0$. Dimensions are in meters.

So the resulting wave is

$$f_3(x,t) = 0.08 \sin[68.5x - 4t] \cos[1.5x + t]$$

In Figs. 12-8 and 12-9, we show the two individual waves prior to interference, while in Fig. 12-10 we show the wave that results from interference at time $t = 0$. Notice that the two waves, when added together, produce a region where they completely cancel at $x = 1$ m.

Energy and Power in Waves

To find the kinetic energy of a wave, we again consider a wave traveling along a string. In this section, we call the kinetic energy K so we don't confuse it with

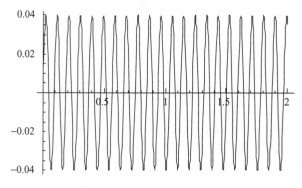

Fig. 12-9 A plot of $f_2(x,t) = 0.04 \sin(67x - 5t)$ at $t = 0$.

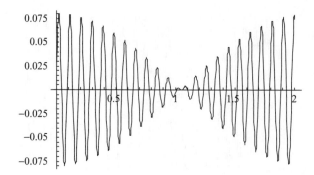

Fig. 12-10 A plot of the superposition $f_3(x,t) = 0.08 \sin[68.5x - 4t] \cos[1.5x + t]$ at $t = 0$, showing how in certain regions the waves add together producing a wave of higher amplitude, but that there is a region where the two waves cancel.

the tension that may be in the string. The kinetic energy of some point P on the string which consists of a small mass element dm is given by (see Fig. 12-1)

$$dK = \frac{1}{2}dm v_y^2$$

The velocity in this equation v_y is the vertical or y-directed velocity of the point P. In (12.3), we stated that this is the derivative of the function describing the wave with respect to time. If we consider the special case of a sinusoidal wave $f(x,t) = A \sin(kx - \omega t)$ then

$$v_y = -\omega A \cos(kx - \omega t)$$

A differential mass element in the string is equal to the mass density σ multiplied by a small length dx of string. Therefore we can write the differential of kinetic energy as

$$dK = \frac{\sigma}{2}dx \omega^2 A^2 \cos^2(kx - \omega t)$$

Dividing by dt we have

$$\frac{dK}{dt} = \frac{\sigma}{2}\frac{dx}{dt}\omega^2 A^2 \cos^2(kx - \omega t)$$

To find the average rate at which the wave carries energy, we can take the time average of this expression. First, we note that the time average of the square of a sinusoidal is

$$\langle \cos^2(kx - \omega t)\rangle = \frac{1}{2}$$

And so the time-averaged rate at which the wave carries kinetic energy is

$$\left\langle \frac{dK}{dt} \right\rangle = \frac{\sigma v}{4} \omega^2 A^2 \qquad (12.19)$$

It can be shown that the time average of the potential energy for a wave on a string is in fact the same

$$\langle V \rangle = \frac{\sigma v}{4} \omega^2 A^2 \qquad (12.20)$$

The average power is the sum of the kinetic and potential energies, which is

$$\langle P \rangle = \frac{1}{2} \sigma v \omega^2 A^2 \qquad (12.21)$$

EXAMPLE 12-5
A string hangs over a pulley as shown in Fig. 12-11. The string is 2 m long and has a mass $m = 7$ g. The mass hanging at the end of the pulley is $M = 1.7$ kg. Suppose that a wave is initiated at the point P that has an amplitude $A = 0.01$ m and a frequency $v = 20$ Hz. How long does it take the disturbance to travel down to the mass M? What is the average power carried by the wave? The system is in equilibrium.

SOLUTION 12-5
The system is in equilibrium therefore the tension in the string is equal to the weight of the mass

$$T = mg = (1.7 \text{ kg})(9.81 \text{ m/s}^2) = 16.7 \text{ N}$$

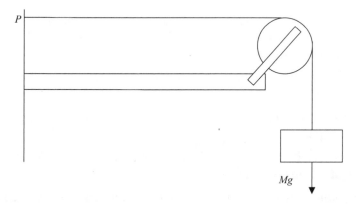

Fig. 12-11   A wave travels down a string over a pulley to a hanging mass.

The density of the string is

$$\sigma = \frac{m}{l} = \frac{7}{2} = 3.5 \text{ g/m} = 0.0035 \text{ kg/m}$$

The velocity of the disturbance is

$$v = \sqrt{\frac{T}{\sigma}} = \sqrt{\frac{16.7}{0.0035}} = 69 \text{ m/s}$$

The time required for the disturbance to travel the length of the string is

$$t = \frac{2 \text{ m}}{69 \text{ m/s}} = 0.029 \text{ s} = 29 \text{ ms}$$

The average power carried by the disturbance is

$$\langle P \rangle = \frac{1}{2}\sigma v \omega^2 A^2 = \frac{1}{2}(0.0035)(69)(2\pi(20))^2 (0.01)^2 = 0.19 \text{ W}$$

Vibrations

We now consider an elastic system with masses. We have already had a taste of this kind of system when we considered a mass-spring system, and when you think of vibrations you can think of a mass-spring system. Basically, this involves the small motion of a mass about an equilibrium position—we call this motion a *mechanical vibration.* Many of the concepts already covered in this chapter such as amplitude, period, cycle, and frequency carry over from wave motion. However, we also consider the following concepts.

- *Natural Frequency:* This is the frequency of a vibrating system in the absence of any external forcing. Vibrations that occur in this case are called *free vibrations.*
- *Forced Vibrations:* These are vibrations caused by the influence of an external force.
- *Transient Vibrations:* These are short-term vibrations in the system that die out quickly.
- *Steady State Vibrations:* These are vibrations that continue in a system long after the transient vibrations die out.
- *Resonance:* Resonance occurs when the frequency of a forced vibration matches the natural frequency of the system.

For a mass spring system with mass m and spring constant k the natural frequency is

$$\omega = \sqrt{\frac{k}{m}} \tag{12.22}$$

EXAMPLE 12-6
A mass $m = 2$ kg is hanging down from a spring with spring constant $k = 0.8$ N/m. If the mass is pulled down by 1 cm and released from rest, find the position and velocity of the mass as functions of time and the natural frequency of the system.

SOLUTION 12-6
We take the resting position of the mass to be at the origin of the coordinate system, which we call x (see Fig. 12-12).

A mass spring system with mass m and spring constant k and no external forcing is described by the differential equation

$$\frac{d^2x}{dt^2} + \frac{k}{m}x = 0 \tag{12.23}$$

The solution of this equation is of the form

$$x(t) = A \cos \omega t + B \sin \omega t \tag{12.24}$$

The velocity is the time derivative of this expression

$$v(t) = -\omega A \sin \omega t + \omega B \cos \omega t \tag{12.25}$$

At $t = 0$, we are told that the mass is pulled down 1 cm. Therefore

$$x(0) = -0.01 = A$$

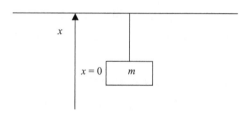

Fig. 12-12   The coordinate system used in Example 12-6.

We are also told that the mass is released *from rest*, therefore the initial velocity satisfies

$$v(0) = 0 = \omega B, \Rightarrow B = 0$$

We conclude that the position of the mass is

$$x(t) = -0.01 \cos \omega t$$

The velocity of the mass is

$$v(t) = 0.01\omega \sin \omega t$$

To completely characterize the solution, we only need to find the natural frequency ω. This is

$$\omega = \sqrt{\frac{k}{m}} = \sqrt{\frac{0.8 \text{ N/m}}{2 \text{ kg}}} = \sqrt{\frac{0.8 \, (\text{kg-m/s}^2)/\text{m}}{2 \text{ kg}}} = 0.63 \text{ Hz}$$

For a system carrying a weight W, the *frequency of vibration* is

$$f = \frac{1}{2\pi} \sqrt{\frac{kg}{W}} \text{ cps} \tag{12.26}$$

where cps is *cycles per second*. Here k is the spring constant and g is the acceleration due to gravity. The period of vibration is found by inverting this quantity

$$T = \frac{1}{f} \tag{12.27}$$

and is measured in seconds.

EXAMPLE 12-7
A 50 lb weight is placed on a spring with a spring constant 11 lb/in. Find the frequency and period of vibration.

SOLUTION 12-7
First, we convert the spring constant so that the units match the acceleration of gravity

$$k = (11 \text{ lb/in}) (12 \text{ in/ft}) = 132 \text{ lb/ft}$$

The frequency of vibration is

$$f = \frac{1}{2\pi}\sqrt{\frac{kg}{W}} = \frac{1}{2\pi}\sqrt{\frac{(132 \text{ lb/ft})(32.2 \text{ ft/s}^2)}{50 \text{ lb}}} = 1.5 \text{ cps}$$

The period of vibration is

$$T = \frac{1}{f} = \frac{1}{1.5} = 0.67 \text{ s}$$

EXAMPLE 12-8
A barrel with radius R is sitting upright in a pool of oil with mass density ρ. The barrel is pushed straight down into the oil by a distance x. What is the natural frequency of oscillation for the barrel if we ignore the damping effects of the oil?

SOLUTION 12-8
The cross-sectional area of the barrel is πR^2. A height x of the barrel, which is just a cylinder, has a volume equal to $\pi R^2 x$. The amount of oil displaced by the barrel is equal to this volume. If the barrel is pushed down, it will be acted on by the buoyant force of this amount of liquid, which is the volume of a cylinder of the liquid multiplied by the mass density. The total mass of oil displaced by the barrel is

$$m = \rho\,(\pi R^2)x$$

The only force acting is gravity, and since the barrel is pushed down we add a negative sign

$$F = ma = -\rho\,(\pi R^2)xg$$

By Newton's second law, this is equal to

$$\frac{dp}{dt} = m\frac{dv}{dt} = m\frac{d^2x}{dt^2}$$

Therefore, we have the following differential equation

$$-\rho\,(\pi R^2)xg = m\frac{d^2x}{dt^2}$$

Dividing by m and then moving all terms to one side, we obtain an equation for a simple harmonic oscillator

$$\frac{d^2x}{dt^2} + \frac{\rho\,(\pi R^2)\,g}{m}x = 0$$

The equation for a simple harmonic oscillator is

$$\frac{d^2x}{dt^2} + \omega^2 x = 0$$

Therefore we deduce that the natural frequency of this system is

$$\omega = \sqrt{\frac{\rho\,(\pi R^2)\,g}{m}}\ \text{rad/s}$$

The frequency of oscillation is

$$f = \frac{1}{2\pi}\omega = \frac{1}{2\pi}\sqrt{\frac{\rho\,(\pi R^2)\,g}{m}}$$

EXAMPLE 12-9

A mass spring system with a mass $m = 2$ kg and spring constant $k = 3000$ N/m is subjected to a force

$$F = 3\cos\omega_f t\ \text{N}$$

Compare the amplitude of vibration if $\omega_f = 20$ rad/s and if $\omega_f = 38$ rad/s.

SOLUTION 12-9

The differential equation for this system can be found by applying Newton's second law

$$m\frac{d^2x}{dt^2} = -kx + 3\cos\omega_f t$$

Rearranging terms

$$\frac{d^2x}{dt^2} + \frac{k}{m}x = \frac{3}{m}\cos\omega_f t$$

The solution to this equation is

$$x(t) = A\cos\omega_f t$$

The derivatives of this function are

$$\frac{dx}{dt} = -\omega_f A \sin \omega_f t$$

$$\frac{d^2x}{dt^2} = -\omega_f^2 A \cos \omega_f t$$

Inserting this into the differential equation, we find the following

$$-\omega_f^2 A \cos \omega_f t + \frac{k}{m} A \cos \omega_f t = \frac{3}{m} \cos \omega_f t$$

Canceling the common factor $\cos \omega_f t$, we find an equation for the amplitude of vibration

$$-\omega_f^2 A + \frac{k}{m} A = \frac{3}{m}$$

$$\Rightarrow A = \frac{3}{m \left(\dfrac{k}{m} - \omega_f^2 \right)}$$

For $k = 3000$ N/m, $m = 2$ kg and $\omega_f = 20$ rad/s

$$A = \frac{3}{2 \left(\dfrac{1500}{2} - (20)^2 \right)} = 1.4 \times 10^{-3} \text{ m}$$

With $\omega_f = 38$ rad/s

$$A_2 = \frac{3}{2 \left(\dfrac{1500}{2} - (38)^2 \right)} = 0.03 \text{ m}$$

The ratio of the forcing frequencies is

$$\frac{38}{20} = 1.9$$

But the ratio of the amplitudes of oscillation is 10 times as large

$$\frac{A_2}{A_1} = \frac{0.03}{1.4 \times 10^{-3}} = 19$$

Quiz

1. A wave is described by $f(x,t) = 5\sin(0.2x - 400t)$. What are the amplitude, wavelength, and period of the wave?

2. A hypothetical wave has a dispersion relation described by $\omega(k) = \alpha k^2 + \beta k$. What are the phase and group velocities of the wave?

3. The tension in a rope is 19 N and the mass density is 8 g/m. What is the velocity of a wave traveling in the rope?

4. A 400 lb machine rests on 4 springs each with spring constant 15 lb/in. What is the period of oscillation?

5. A mass spring system with $m = 7$ kg and $k = 7000$ N/m is subjected to a force $F = 7\cos\omega_f t$ N. What is the amplitude of oscillation if $\omega_f = 84$ rad/s?

CHAPTER 13

Advanced Mechanics

In Chapter 11, we spent some time talking about energy and energy conservation. In this chapter, we expand our look at energy by considering methods of doing dynamics that are based on energy rather than considering forces. These methods of doing mechanics were developed in the 18th and 19th centuries long after Newton had proposed his three laws. This is a more sophisticated approach, but it is in some sense more fundamental and insightful, and can simplify some problems. We begin by considering a new quantity called the Lagrangian, named after its originator Joseph-Louis Lagrange, who lived from 1736 to 1813. His methods were laid out in a book on mechanics he wrote at the age of 19. One reason the book was famous is that it did not contain a single diagram.

The Lagrangian

We now introduce a new scalar quantity that is based on the energy of a system. Following the notation introduced in Chapter 11, we let T denote kinetic energy and V denote potential energy. The Lagrangian L is defined as

$$L = T - V \qquad (13.1)$$

As you can guess from your studies on energy, the Lagrangian can be a function of position, velocity, and time. The Lagrangian can be used to derive the equations of motion for a system. This is done by calculating the *Euler–Lagrange* equations. For simplicity, first we consider one-dimensional motion. The Euler–Lagrange equations take the form

$$\frac{d}{dt}\frac{\partial L}{\partial \dot{x}} = \frac{\partial L}{\partial x} \tag{13.2}$$

Remember that $\dot{x} = dx/dt = v$, the velocity of the particle.

EXAMPLE 13-1
Determine the equations of motion for a mass–spring system, using the Euler–Lagrange equation.

SOLUTION 13-1
We begin by writing down the kinetic and potential energy for this system. First, we know the kinetic energy is simply

$$T = \frac{1}{2}mv^2$$

Noting the form of (13.2), we rewrite this in terms of the first time derivative of position

$$T = \frac{1}{2}m\dot{x}^2$$

In terms of the spring constant k, the potential energy of a mass–spring system is

$$V = \frac{1}{2}kx^2$$

Therefore, the Lagrangian for a mass–spring system is

$$L = T - V = \frac{1}{2}m\dot{x}^2 - \frac{1}{2}kx^2$$

Now let's compute the quantities we need. First, we take the derivative of the Lagrangian with respect to position

$$\frac{\partial L}{\partial x} = \frac{\partial}{\partial x}\left(\frac{1}{2}m\dot{x}^2 - \frac{1}{2}kx^2\right) = -kx$$

Next, we calculate the derivative of the Lagrangian with respect to $\dot{x}$, giving

$$\frac{\partial L}{\partial \dot{x}} = \frac{\partial}{\partial \dot{x}} \left(\frac{1}{2}m\dot{x}^2 - \frac{1}{2}kx^2 \right) = m\dot{x}$$

Looking at (13.2), we see that we need to take the time derivative of this quantity

$$\frac{d}{dt}\frac{\partial L}{\partial \dot{x}} = \frac{d}{dt}(m\dot{x}) = m\ddot{x}$$

Now we equate the terms we have to give the equations of motion

$$m\ddot{x} = -kx$$

Or, moving all terms to one side

$$m\ddot{x} + kx = 0$$

In the first example we derived the equation for a simple harmonic oscillator by looking at the energy in the system. This result is not particularly illuminating since the equations of motion are easy to write down. The utility of the Lagrangian methods is that there will be times when it's a simple matter to write down the energy, but it's not so simple to write down the equations of motion. By calculating the Lagrangian and the Euler–Lagrange equations, however, the equations of motion can be obtained.

EXAMPLE 13-2
In this example, we use Lagrangian methods to derive some of the equations of kinematics used in Chapter 8. Find the equations of motion for a particle in free fall from a height y.

SOLUTION 13-2
The kinetic energy of the particle is just $T = \frac{1}{2}m\dot{y}^2$. The potential energy is

$$V = mgy$$

The Lagrangian is

$$L = T - V = \frac{1}{2}m\dot{y}^2 - mgy$$

Using a procedure identical to that in the last example, we find

$$\frac{d}{dt}\frac{\partial L}{\partial \dot{y}} = \frac{d}{dt}(m\dot{y}) = m\ddot{y}$$

Now,

$$\frac{\partial L}{\partial y} = -mg$$

and so the equation of motion for this system is

$$m\ddot{y} = -mg, \Rightarrow \ddot{y} = -g$$

Looking at the equation, recall that acceleration is the second derivative of position with respect to time. This equation tells us that the acceleration of the particle is precisely the gravitational acceleration. Now let's integrate to find the velocity. We call the initial velocity, which is the constant of integration, v_0.

$$\dot{y} = \int g\, dt = -gt + v_0$$

Since $\dot{y} = v$, notice that we have obtained equation (8.12), that is $v = v_0 - gt$. We can integrate a second time to obtain an equation for position. We call the constant of integration y_0

$$y = \int (-gt + v_0)\, dt = -\frac{g}{2}t^2 + v_0 t + y_0$$

Rearranging terms, we have

$$y - y_0 = v_0 t - \frac{1}{2}gt^2$$

This is just equation (8.13).

EXAMPLE 13-3
A particle is moving in the potential $V(x) = \alpha x^3$ where α is a constant. Find the equations of motion for this system.

SOLUTION 13-3
The kinetic energy is $T = \frac{1}{2}m\dot{x}^2$. The Lagrangian is

$$L = T - V = \frac{1}{2}m\dot{x}^2 - \alpha x^3$$

Once again,

$$\frac{d}{dt}\frac{\partial L}{\partial \dot{x}} = \frac{d}{dt}(m\dot{x}) = m\ddot{x}$$

We also find that

$$\frac{\partial L}{\partial x} = -3\alpha x^2$$

Writing the Euler–Lagrange equations as

$$\frac{d}{dt}\frac{\partial L}{\partial \dot{x}} - \frac{\partial L}{\partial x} = 0$$

We find that the equations of motion for this system are

$$m\ddot{x} + 3\alpha x^2 = 0$$

Systems in Multiple Dimensions

When a particle is moving in more than one dimension, the Euler–Lagrange equations constitute a set. To each independent coordinate, there will correspond an Euler–Lagrange equation. In three dimensions, if we symbolically denote each coordinate by a subscript x_i, where $i = 1, 2, 3$ then we can write the Euler–Lagrange equations as

$$\frac{d}{dt}\frac{\partial L}{\partial \dot{x}_i} - \frac{\partial L}{\partial x_i} = 0 \tag{13.3}$$

Confused? Let's make things worse with an example.

EXAMPLE 13-4
Derive the equations of motion for a projectile with no air resistance.

SOLUTION 13-4
This is a standard example of two-dimensional motion, and we analyzed projectile motion in Chapters 8 and 9. First, we compute the kinetic energy of the system. Using the standard Cartesian coordinates in the plane, there will be motion in both the x and y directions. Therefore the kinetic energy is the sum of the kinetic energies in each direction, i.e.,

$$T = \frac{1}{2}m\dot{x}^2 + \frac{1}{2}m\dot{y}^2$$

The potential energy acting in this problem is due to gravity. At a height y, the potential energy of the projectile is

$$V = mgy$$

The Lagrangian is therefore given by

$$L = T - V = \frac{1}{2}m\dot{x}^2 + \frac{1}{2}m\dot{y}^2 - mgy$$

To derive the equations of motion, we consider the x and y components separately, writing down the Euler–Lagrange equation for each. First, we consider x. We have

$$\frac{\partial L}{\partial x} = \frac{\partial}{\partial x}\left(\frac{1}{2}m\dot{x}^2 + \frac{1}{2}m\dot{y}^2 - mgy\right) = 0$$

Next,

$$\frac{\partial L}{\partial \dot{x}} = \frac{\partial}{\partial \dot{x}}\left(\frac{1}{2}m\dot{x}^2 + \frac{1}{2}m\dot{y}^2 - mgy\right) = m\dot{x}$$

And so

$$\frac{d}{dt}\frac{\partial L}{\partial \dot{x}} = m\ddot{x}$$

Application of (13.3) for the x coordinate gives us

$$\ddot{x} = 0$$

where we were able to throw away the mass since we have a single term set equal to zero, so it cancels. Integrating once gives us the velocity of the projectile in the x direction

$$\dot{x} = v_x = v_{0x}$$

We have found that the velocity is a constant, consistent with our earlier work on this problem. Integrating again to obtain the position

$$x(t) = v_{0x}t + x_0$$

Now we turn our attention to the y coordinate. First, notice that

$$\frac{\partial L}{\partial y} = \frac{\partial}{\partial y}\left(\frac{1}{2}m\dot{x}^2 + \frac{1}{2}m\dot{y}^2 - mgy\right) = -mg$$

Next, we find that

$$\frac{\partial L}{\partial \dot{y}} = \frac{\partial}{\partial \dot{y}}\left(\frac{1}{2}m\dot{x}^2 + \frac{1}{2}m\dot{y}^2 - mgy\right) = m\dot{y}$$

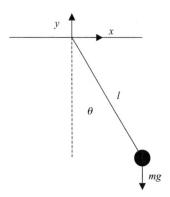

Fig. 13-1   A plane pendulum.

So that

$$\frac{d}{dt}\left(\frac{\partial L}{\partial \dot{y}}\right) = m\ddot{y}$$

Application of (13.3) to the y coordinate gives us the Euler–Lagrange equation

$$m\ddot{y} + mg = 0, \Rightarrow \ddot{y} + g = 0$$

This is the same equation we saw in Example 13-2, so integration will give us the equations of free fall.

EXAMPLE 13-5
Find the equations of motion for a plane pendulum of length l.

SOLUTION 13-5
The situation is shown in Fig. 13-1.
   We can analyze this problem in the following way. First, note that plane polar coordinates in this case are related to Cartesian coordinates via the relations

$$x = r\sin\theta$$
$$y = -r\cos\theta$$

(look at Fig. 13-1 and notice where we put the origin of the coordinate system). In this problem, the coordinate r is fixed, in fact it is the length of the pendulum. Therefore

$$x = l\sin\theta$$
$$y = -l\cos\theta$$

The kinetic energy is similar to that in the projectile motion example. That is

$$T = \frac{1}{2}m\dot{x}^2 + \frac{1}{2}m\dot{y}^2$$

However in this problem, our concern is the change of the angle θ with time. Using the chain rule, we find

$$\dot{x} = \frac{dy}{dt} = \frac{d}{dt}(l\sin\theta) = l\frac{d\theta}{dt}\cos\theta = l\dot{\theta}\cos\theta$$

$$\dot{y} = \frac{dx}{dt} = \frac{d}{dt}(-l\cos\theta) = l\frac{d\theta}{dt}\sin\theta = l\dot{\theta}\sin\theta$$

Now we can write the kinetic energy of the system as

$$T = \frac{1}{2}m\dot{x}^2 + \frac{1}{2}m\dot{y}^2$$
$$= \frac{1}{2}ml^2\dot{\theta}^2\cos^2\theta + \frac{1}{2}ml^2\dot{\theta}^2\sin^2\theta = \frac{1}{2}ml^2\dot{\theta}^2(\cos^2\theta + \sin^2\theta) = \frac{1}{2}ml^2\dot{\theta}^2$$

The potential energy is just the gravitational potential energy for the pendulum bob at height y. This is

$$V = mgy = -mgl\cos\theta$$

This allows us to write down the Lagrangian for the system

$$L = T - V = \frac{1}{2}ml^2\dot{\theta}^2 + mgl\cos\theta$$

Now we proceed to find an equation of motion for the angular coordinate. First, we find

$$\frac{\partial L}{\partial \theta} = \frac{\partial}{\partial \theta}\left(\frac{1}{2}ml^2\dot{\theta}^2 + mgl\cos\theta\right) = -mgl\sin\theta$$

Next,

$$\frac{\partial L}{\partial \dot{\theta}} = \frac{\partial}{\partial \dot{\theta}}\left(\frac{1}{2}ml^2\dot{\theta}^2 + mgl\cos\theta\right) = ml^2\dot{\theta}$$

And so

$$\frac{d}{dt}\frac{\partial L}{\partial \dot\theta} = ml^2\ddot\theta$$

The Euler–Lagrange equation gives us the equation of motion

$$\frac{d}{dt}\frac{\partial L}{\partial \dot\theta} - \frac{\partial L}{\partial \theta} = ml^2\ddot\theta + mgl\sin\theta = 0$$

Canceling the mass and dividing by l gives us the pendulum equation

$$\ddot\theta + \frac{g}{l}\sin\theta = 0$$

EXAMPLE 13-6
Now consider a driven plane pendulum. The pendulum is attached to an oscillating support that oscillates at frequency ω with amplitude h. Find the equation of motion for the angle the pendulum makes with the vertical.

SOLUTION 13-6
The situation is shown in Fig. 13-2.
   This example is similar to the last one, except the y coordinate in this problem is lengthened by an amount determined by $h(t) = h\cos\omega t$. We have

$$x = l\sin\theta$$
$$y = -l\cos\theta - h\cos\omega t$$

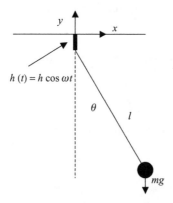

Fig. 13-2   A driven pendulum.

Then

$$\dot{x} = \frac{dx}{dt} = \frac{d}{dt}(l\sin\theta) = l\frac{d\theta}{dt}\cos\theta = l\dot{\theta}\cos\theta$$

$$\dot{y} = \frac{dy}{dt} = \frac{d}{dt}(-l\cos\theta - h\cos\omega t) = l\frac{d\theta}{dt}\sin\theta + \omega h\sin\omega t$$

$$= l\dot{\theta}\sin\theta + \omega h\sin\omega t$$

Therefore

$$T = \frac{1}{2}m\dot{x}^2 + \frac{1}{2}m\dot{y}^2$$

$$= \frac{1}{2}ml^2\dot{\theta}^2\cos^2\theta + \frac{1}{2}ml^2\dot{\theta}^2\sin^2\theta + \frac{1}{2}m\omega^2 h^2\sin^2\omega t + ml\omega h\dot{\theta}\sin\theta\sin\omega t$$

$$= \frac{1}{2}ml^2\dot{\theta}^2\left(\cos^2\theta + \sin^2\theta\right) + \frac{1}{2}m\omega^2 h^2\sin^2\omega t + ml\omega h\dot{\theta}\sin\theta\sin\omega t$$

$$= \frac{1}{2}ml^2\dot{\theta}^2 + \frac{1}{2}m\omega^2 h^2\sin^2\omega t + ml\omega h\dot{\theta}\sin\theta\sin\omega t$$

The potential energy is again mgy. In this case

$$V = -mgy = -mg(l\cos\theta + h\cos\omega t)$$

The Lagrangian is

$$L = T - V$$

$$= \frac{1}{2}ml^2\dot{\theta}^2 + \frac{1}{2}m\omega^2 h^2\sin^2\omega t + ml\omega h\dot{\theta}\sin\theta\sin\omega t + mg(l\cos\theta + h\cos\omega t)$$

Now we calculate the terms required for the Euler–Lagrange equation. First, we have

$$\frac{\partial L}{\partial\dot{\theta}} = ml^2\dot{\theta} + ml\omega h\sin\theta\sin\omega t$$

Therefore

$$\frac{d}{dt}\frac{\partial L}{\partial\dot{\theta}} = ml^2\ddot{\theta} + ml\omega^2 h\sin\theta\cos\omega t + ml\omega h\dot{\theta}\cos\theta\sin\omega t$$

And

$$\frac{\partial L}{\partial\theta} = ml\omega h\dot{\theta}\cos\theta\sin\omega t - mgl\sin\theta$$

The Euler–Lagrange equation (13.3) gives

$$ml^2\ddot{\theta} + ml\omega^2 h \sin\theta \cos\omega t + ml\omega h\dot{\theta}\cos\theta \sin\omega t - ml\omega h\dot{\theta}\cos\theta \sin\omega t$$

$$+ mgl\sin\theta = ml^2\ddot{\theta} + ml\omega^2 h\sin\theta\cos\omega t + mgl\sin\theta = 0$$

Canceling the mass gives

$$l^2\ddot{\theta} + l\omega^2 h\sin\theta\cos\omega t + gl\sin\theta = 0$$

Now we divide by the length squared (the leading coefficient)

$$\ddot{\theta} + \frac{\omega^2 h\sin\theta}{l}\cos\omega t + g\sin\theta = 0$$

The Hamiltonian

It turns out we can go a step further in the analysis of a system via energy for systems where the potential energy does not depend on velocity. We now consider the total energy of a system as expressed by a quantity called the *Hamiltonian*. While the Lagrangian is a function of position, velocity, and time, the Hamiltonian is a function of position, *momentum*, and time. To begin, we need to use a formal procedure to obtain the momentum of a system which is described by a Lagrangian. This is done by calculating the derivative of the Lagrangian with respect to velocity

$$p = \frac{\partial L}{\partial \dot{x}} \tag{13.4}$$

The Hamiltonian H can be constructed by writing

$$H = p\dot{x} - L \tag{13.5}$$

If motion occurs in more than one dimension, we can label the coordinates with a subscript and then

$$p_i = \frac{\partial L}{\partial \dot{x}_i}$$

And then the Hamiltonian is given by

$$H = \sum_i p_i\dot{x}_i - L \tag{13.6}$$

Once the Hamiltonian is in hand, the problem can be solved using Hamilton's equations of motion

$$\dot{x}_i = \frac{\partial H}{\partial p_i} \tag{13.7}$$

$$\dot{p}_i = -\frac{\partial H}{\partial x_i} \tag{13.8}$$

While the Hamiltonian method seems like mathematical overkill, it turns out to be extremely useful in many situations. For example, in atomic and subatomic physics the Hamiltonian is extremely important, and plays a central role in quantum theory. To illustrate the techniques, we will redo an example.

EXAMPLE 13-7
Redo Example 13-4, where the equations of motion were found for projectile motion by considering the Hamiltonian.

SOLUTION 13-7
In Example 13-4, we found that the Lagrangian is

$$L = T - V = \frac{1}{2}m\dot{x}^2 + \frac{1}{2}m\dot{y}^2 - mgy$$

With two coordinates, we need to find two momenta. The momentum conjugate to the x coordinate is

$$p_x = \frac{\partial L}{\partial \dot{x}} = \frac{\partial}{\partial \dot{x}}\left(\frac{1}{2}m\dot{x}^2 + \frac{1}{2}m\dot{y}^2 - mgy\right) = m\dot{x}$$

The momentum conjugate to the y coordinate is

$$p_y = \frac{\partial L}{\partial \dot{y}} = \frac{\partial}{\partial \dot{y}}\left(\frac{1}{2}m\dot{x}^2 + \frac{1}{2}m\dot{y}^2 - mgy\right) = m\dot{y}$$

We can then rewrite $\dot{x}$ and $\dot{y}$ in terms of the momentum conjugate to each coordinate.

$$\dot{x} = \frac{p_x}{m}$$

$$\dot{y} = \frac{p_y}{m}$$

This allows us to rewrite the Lagrangian in terms of momentum. Using these substitutions

$$L = \frac{p_x^2}{2m} + \frac{p_y^2}{2m} - mgy$$

Now we have

$$\sum_i p_i \dot{x}_i = p_x \dot{x} + p_y \dot{y} = \frac{p_x^2}{m} + \frac{p_y^2}{m}$$

The Hamiltonian can be written down using (13.6). We find

$$H = \sum_i p_i \dot{x}_i - L$$

$$= \frac{p_x^2}{m} + \frac{p_y^2}{m} - \left(\frac{p_x^2}{2m} + \frac{p_y^2}{2m} - mgy \right)$$

$$= \frac{p_x^2}{2m} + \frac{p_y^2}{2m} + mgy$$

An important point to notice is that the Hamiltonian we've derived is the sum of the kinetic and potential energies of the system. Now we can use Hamilton's equations (13.7) and (13.8) to write down the equations of motion. Considering the x coordinate first, we have

$$\frac{dx}{dt} = \frac{\partial H}{\partial p_x} = \frac{\partial}{\partial p_x} \left(\frac{p_x^2}{2m} + \frac{p_y^2}{2m} + mgy \right) = \frac{p_x}{m}$$

Next, we have

$$\frac{dp_x}{dt} = -\frac{\partial H}{\partial x} = -\frac{\partial}{\partial x} \left(\frac{p_x^2}{2m} + \frac{p_y^2}{2m} + mgy \right) = 0$$

This tells us the momentum in the x direction is a *constant*. Therefore

$$\frac{dx}{dt} = \frac{p_x}{m} \Rightarrow x(t) = \frac{p_x}{m} t + x_0$$

Now for the y direction

$$\frac{dy}{dt} = \frac{\partial H}{\partial p_y} = \frac{\partial}{\partial p_y} \left(\frac{p_x^2}{2m} + \frac{p_y^2}{2m} + mgy \right) = \frac{p_y}{m}$$

And

$$\frac{dp_y}{dt} = -\frac{\partial H}{\partial y} = -\frac{\partial}{\partial y}\left(\frac{p_x^2}{2m} + \frac{p_y^2}{2m} + mgy\right) = -mg$$

Hence

$$\frac{dp_y}{dt} = -mg \Rightarrow p_y = -mgt + p_{0y}$$

Therefore the y coordinate satisfies

$$\frac{dy}{dt} = \frac{p_y}{m} = -gt + \frac{p_{0y}}{m}, \Rightarrow$$

$$y(t) = -\frac{g}{2}t^2 + \frac{p_{0y}}{m}t + y_0$$

Quiz

1. A particle is moving in a one-dimensional potential described by
   $V(x) = x^2 - x$. Write down the equations of motion.

2. Suppose that a particle is constrained to move in the plane where the
   potential is $V = r\theta$. Write down the Lagrangian for this system.

3. What are the equations of motion for problem 2?

The following data is for questions 4, 5, and 6. A bead is constrained to
move on a hoop with radius R rotating with angular frequency Ω. The
position of the bead is given by

$$x = R\sin\theta\cos\Omega t$$
$$y = R\sin\theta\sin\Omega t$$
$$z = -R\cos\theta$$

The potential energy of the bead is $V = mgR(1 - \cos\theta)$

4. What is the kinetic energy of the bead?

5. Write down the Lagrangian.

6. Find the equation of motion for the bead.

7. Duplicate the results of Example 13-5 by using Hamiltonian methods.

Final Exam

Let $\vec{A} = 2\hat{x} + 2\hat{y} - \hat{z}$ and $\vec{B} = 2\hat{x} + 2\hat{y}$.

1. What is the length of A?
2. What is the length of B?
3. What is $\vec{A} \cdot \vec{B}$?
4. If $\vec{A} = \hat{x} - 5\hat{y} + 2\hat{z}$ and $\vec{B} = \hat{x} + \hat{y} + \hat{z}$, find $\vec{A} + \vec{B}$
5. Find ∇T if $T = x \cos y$
6. What is the total mass of a cylinder with height 6 m and radius 2 m and a mass density of 2 kg/m^3?
7. A particle with mass $m = 2$ kg is moving with an initial velocity $v = 1$ m/s. It starts from the origin, under the influence of a constant force $F = 7$ N. What are the position and velocity of the particle after 10 s?
8. What is the vector equation that defines the moment of a force?
9. A force with a magnitude of 221 N is directed from the origin to (5,1,3). What are the x, y, and z components of the force?
10. What is the x component of the moment for the force in problem 9?

11. A body is acted on by forces of 20 N, −30 N, and 17 N in the plane. What is the resultant?

12. What is the direction of the resultant in problem 11?

13. What is the formula that determines the action line of a resultant?

14. If the following forces act on a body, what is the magnitude of the resultant?

$$\mathbf{F}_1 = 11\mathbf{i} + 2\mathbf{j}$$
$$\mathbf{F}_2 = 2\mathbf{i} + 4\mathbf{j}$$
$$\mathbf{F}_3 = 20\mathbf{i} + 14\mathbf{j}$$

15. Forces of 20 N, −10 N, and 3 N act on a beam at distances of 2 m, 5 m, and 20 m, respectively. What is the resultant?

16. Forces of 20 N, −10 N, and 3 N act on a beam at distances of 2 m, 5 m, and 20 m, respectively. What is the moment?

17. Forces of 20 N, −10 N, and 3 N act on a beam at distances of 2 m, 5 m, and 20 m, respectively. Where is the action line of the resultant?

18. What is the condition on forces for static equilibrium?

19. What is the condition on moments for static equilibrium?

20. The mass of a man who is 5 ft 4 in tall is 6.1 slugs. Would his doctor tell him to loose weight?

21. What is the force of gravity between a rock weighing 50 kg and a rock weighing 300 kg that are 2 m apart?

22. A satellite takes 118 min to orbit the earth. What height is it above the earth's surface?

23. In a strange star system, a mass has a gravitational acceleration described by $\mathbf{g} = -\ GMe^{-\alpha r}\hat{\mathbf{r}}$ where α is a constant. What is the gravitational potential?

24. The gravitational potential is $\phi = \dfrac{2GM \cos r}{r^3}$. What is the gravitational field vector?

25. What is the formula for escape velocity?

26. What is the moment of a mass m displaced a distance x from the fulcrum?

27. What is the condition for balance with two masses placed on a beam?

28. If several masses are placed on a beam, where is the center of mass?

29. Four masses $m_1 = 2$ kg, $m_2 = 1$ kg, $m_3 = 3$ kg, and $m_4 = 5$ kg are located at the positions $x_1 = 1$ m, $x_2 = 3$ m, $x_3 = 3$ m, and $x_4 = 10$ m, respectively. Where is the center of mass located for this system?

30. What is the moment of inertia for a continuous system?

31. Define the radius of gyration.

32. A suspension bridge with a span of 211 m is supported by two cables, each of which carries a load of 13000 N/m. If the sag of each cable is 18 m, what is the tension at the midpoint of the cable?

33. Each cable in a suspension bridge with a 400 ft span and a 22 ft sag carries a load of 720 lb/ft. What is the tension at the midpoint of the cable and what is the tension at the supports found on each end?

34. A cable spans 200 ft and sags by 34 ft. How long is the cable?

35. What function most closely describes the curve known as a catenary?

36. The load in a parabolic cable is supported in what direction?

37. A catenary cable is described by $c = 118$ ft, has maximum tension of 550 lb, and carries a load equal to 3 lb/ft. What is the sag?

38. A catenary cable is described by $c = 118$ ft, has maximum tension of 550 lb, and carries a load equal to 3 lb/ft. What is the tension at the midpoint of the cable?

39. With respect to the direction of motion, where does the frictional force point?

40. What does static friction resist?

41. What is the formula for the normal force?

42. Given a normal force N, what is the frictional force?

43. A belt is about to slip. How are the tensions on the two sides of the belt related?

44. A belt is wrapped around a pulley. On the slack side, the tension is 350 N while the tension in the tight side is 400 N. Find the coefficient of friction between the belt and the pulley.

45. A cart has wheels that are 16 in in diameter. A wheel carrying a load of 350 lb is stuck in a depression, and it is found that a force of 210 lb is required to roll the cart out. What is the coefficient of rolling resistance?

46. Define acceleration.

47. A particle moves with a constant acceleration 2 m/s². If the initial velocity is 1 m/s, how long does it take the particle to reach a speed of 6 m/s?

48. A rocket sled is moving along a straight track. Its position is known to vary with time according to $x(t) = e^{-2t}(180t^2 - 50)$ m where time is measured in seconds. Find the velocity of the sled at $t = 1$ s.

49. A red sports car driven by an angry woman is speeding down a rural highway. Fearing the impending presence of a police officer, she breaks to slow from 100 miles per hour to 45 miles per hour over a distance of 375 ft. How much time does it take for the car to slow to 45 miles per hour?

50. Find the acceleration in the previous problem, assuming that it's constant.

51. A man looking out from a tall building drops his keys. They strike the ground with a speed $v = 274$ ft/s. How high is the lookout tower?

52. How long did it take for the keys to reach the ground?

53. What is the position vector from the origin?

54. The position of a particle is given by $\mathbf{r} = \cos t\, \mathbf{i} - \sin t\, \mathbf{j} + t^2\, \mathbf{k}$. What is the particle's velocity vector?

55. If $\mathbf{r} = \cos t\, \mathbf{i} - \sin t\, \mathbf{j} + t^2\, \mathbf{k}$ what is the particle's acceleration vector?

56. If $\mathbf{r} = \cos 3t\, \mathbf{i} - \sin 3t\, \mathbf{j}$ where position is measured in meters, what is the speed of the particle?

57. What is the curvature for a particle moving on an arbitrary curved path?

58. The position vector for a particle is given by $\mathbf{r} = 3t^2\, \mathbf{i} + 2t\, \mathbf{j} + \mathbf{k}$. What is the velocity vector?

59. If $\mathbf{r} = 3t^2\, \mathbf{i} + 2t\, \mathbf{j} + \mathbf{k}$, what is the unit tangent vector to the curve describing the path of the particle?

60. What is the time to impact for a projectile with muzzle velocity $v_0 = 45$ m/s if the gun is angled at $33°$ with respect to the horizon? Neglect air resistance.

61. How far does a projectile travel if the muzzle velocity $v_0 = 45$ m/s if the gun is angled at $33°$ with respect to the horizon? Neglect air resistance.

62. What is the condition for maximum range for a given muzzle velocity if there is no air resistance?

63. An artillery piece is situated atop a mountain that is 1.2 km above the valley below. If the gun is situated at an angle of $\theta = 37°$ above the horizontal, and the projectile travels 6.7 km, what is the muzzle velocity?

64. An artillery piece is situated atop a mountain that is 1.2 km above the valley below. If the gun is situated at an angle of $\theta = 37°$ above the horizontal, and the projectile travels 6.7 km. What is the time of flight?

65. A man in a NASA super G Force test center is moving about a radius of 8 m with an acceleration of 2 g. What is his speed?

66. A satellite orbits a newly discovered planet hiding behind Jupiter. It orbits at a height of 210 km with a velocity of 8.3 km/s. What is the acceleration of gravity at this height above planet mysterio?

67. The velocity of a 2 kg body is 17 m/s. What is its momentum?

68. What does conservation of momentum mean?

69. How is conservation of momentum expressed in terms of Newton's laws?

70. Describe uniform motion mathematically.

71. What is a Newton in terms of the fundamental units of mass, length, and time?

72. What is the force of kinetic friction?

73. A body is moving in a potential $U = 2r^2 e^{-r}$. What is the force on the particle?

74. A particle which starts at the origin is under the influence of a potential given by $U(x) = \alpha x^2$. What is the velocity of the particle up to a constant?

75. A mass spring system is moving such that the force on the mass is 7 N when the mass is 2 m to the left of the origin. What is the spring constant?

76. What is the fundamental frequency of a mass–spring system with mass m and spring constant k?

77. Consider a mass–spring system. At $t = 0$, the mass is located at $x = 2$ ft and the velocity of the mass is $v(0) = 0$. Determine the kinetic energy of the mass as a function of time.

78. The spring constant for a mass–spring system is 0.2 N/m. What is the potential when the mass is located at $x = 2$ cm?

79. What is the differential equation for a mass–spring system with damping?

80. What is the best way to describe overdamping?

81. How does an underdamped system differ from a critically damped system?

82. In uniform circular motion, in what direction does the acceleration vector point?

83. A flywheel has a radius of 0.45 m. The flywheel is initially at rest, then accelerates to 1800 rpm in 12 s. What is the angular acceleration?

84. A flywheel accelerates from rest to 3200 rpm. If this takes 14 s, how many revolutions does the flywheel make in that time?

85. What is the kinetic energy of rotation?

86. A flywheel with total mass $m = 1450$ kg and radius of gyration 1.1 m is accelerated from rest to 2000 rpm in 75 s. What is the moment?

87. A particle has an angular momentum $\mathbf{L} = 3t^3\mathbf{i} + 7t\,\mathbf{j} - \cos t\,\mathbf{k}$. What is the moment on the particle?

88. A 3 kg mass at time t is being acted upon by a force $\mathbf{F} = 4\,\mathbf{k}$ N. Position is measured in meters and the particle is located at $\mathbf{r} = 2.0\mathbf{i} + 2.0\mathbf{j}$. If the velocity in m/s is given by $\mathbf{v} = 3\mathbf{i} + \mathbf{j}$, what torque is acting on the particle?

89. How many joules are in an electron volt?

90. A car weighing 2500 lb is traveling at 95 miles per hour. What is its kinetic energy?

91. A force of 122 N is applied to move a box horizontally 3 m. How much work was done?

92. If the applied force acts in the same direction as the motion, then the work is
    (a) Negative.
    (b) Positive.
    (c) Forces parallel to motion do no work.

93. What is the efficiency of an engine that puts out 322 Watts if the input power is 480 Watts?

94. The dispersion relation for a certain medium is $\omega(k) = 4k^2$. What is the group velocity?

95. The dispersion relation for a certain medium is $\omega(k) = 9k^2$. What is the phase velocity?

96. The Lagrangian for a system is $L = m\dot{x}^2 - 3x^5$. What is the equation of motion?

97. A particle is moving in a region where the potential is $V = \beta x$. What is the Lagrangian?

98. What are the equations of motion for the potential in problem 97?

99. For some system, the Lagrangian is $L = \frac{1}{2}m\dot{x}^2 + \alpha\dot{x} - \beta x^2$. What is the conjugate momentum?

100. What is the Hamiltonian when $L = \frac{1}{2}m\dot{x}^2 + \alpha\dot{x} - \beta x^2$?

Quiz Solutions and Final Exam Answers

Quiz Solutions

Chapter 1

1. (a) 6.5, 2.8
    (b) 10
    (c) 57°
2. $\vec{A} + \vec{B} = 3\hat{x} + 3\hat{y} + 8\hat{z}$, $\vec{A} - \vec{B} = 5\hat{x} - 7\hat{y} - 6\hat{z}$

$$|A| = \sqrt{21}, \ |B| = \sqrt{77}$$

3. $\vec{A} \cdot \vec{B} = -7$, no
4. $r = \sqrt{21}$, $\theta = 77°$, $\phi = -27°$, $\vec{A} = 3.3\,\hat{r} - 0.3\,\hat{\theta} - 0.9\,\hat{\phi}$

5. $\nabla T (1, 1, 1) \approx 5.2\hat{x} + 2.3\hat{y} - 2.7\hat{z}$, most rapidly increasing in x direction

6. $\dfrac{784\pi}{3}$

7. $\vec{p} = 8mt\,\hat{i} - 3m\hat{j}$

8. $\nabla \times \vec{F} = 0$, is conservative.

9. $\vec{\nabla} T \cdot d\vec{l} = (5y^3 - 2y)dx + (15xy^2 - 2x)dy$

10. Use $d\vec{a} = a^2 \sin\theta d\theta d\phi \, \hat{r}$

Chapter 2

1. 2 m/s
2. 16 m/s
3. $\mathbf{F} = m\mathbf{a}$

4. $F \cos \omega t = m \dfrac{d^2x}{dt^2}$

5. $x(t) = A \cos \left(\sqrt{\dfrac{k}{m}}t \right) + B \sin \left(\sqrt{\dfrac{k}{m}}t \right)$

6. 22.4 N, 44.5 N, 67.3 N

7. -314.8 N-m, -0.1 N-m, 22.5 N-m

Chapter 3

1. 10.6 N
2. 53°
3. 37 N
4. 155 N-m
5. 4.2 m
6. $M = 880$ N

Chapter 4

1. 274 m/s^2
2. 3352 lb
3. 1.93×10^{-8} N
4. 618 km/s

Chapter 5

1. 35 m
2. 2.4 m

3. $(\bar{x}, \bar{y}, \bar{z}) = (7/13, 16/13, 19/13)$

4. $Q_x = \dfrac{ab^2}{4}$, $Q_y = \dfrac{2b^{5/2}\sqrt{a}}{5}$

5. $(10/3, 1/3)$

6. $I_y = \dfrac{m}{2}(3R^2 + h^2)$

7. $I = \dfrac{2}{5}mR^2$

8. $I = \dfrac{3}{5}\left(\dfrac{R^2}{4} + h^2\right)$

9. $k = \dfrac{1}{\sqrt{2}}\sqrt{(3R^2 + h^2)}$

10. $k = \sqrt{\dfrac{3}{5}\left(\dfrac{R^2}{4} + h^2\right)}$

Chapter 6

1. 971,429 lb

2. 28 ft

3. Just about 11 ft

4. $T_{\text{MAX}} = 19{,}688$ N, $H = 19{,}228$ N

Chapter 7

1. 79.8 lb

2. 63.3 N

3. 8.1 m/s^2

4. 28 N

5. 2.2 N

Chapter 8

1. $x(3) = 30$ m, $v(3) = 46$ m/s, $a(3) = 36$ m/s^2

2. $a(t) = 3e^{-t}(4\sin 2t - 3\cos 2t)$

3. 179.2 ft

4. 4.9 s

5. 188 ft

6. 6.8 s

7. 4796 ft

8. 2.0 s, 20.4 m

9. $\mathbf{T} = \dfrac{-2\sin 2t\,\mathbf{i} + 3\cos 3t\,\mathbf{j} + 2t\,\mathbf{k}}{\sqrt{4\sin^2 2t + 9\cos^2 3t + 4t^2}}$

10. $\mathbf{v} = (\cos t - t\sin t)\,\mathbf{i} + (\sin t + t\cos t)\,\mathbf{j} + 2t\,\mathbf{k}$,

$\mathbf{a} = -(2\sin t - t\cos t)\,\mathbf{i} + (2\cos t - t\sin t)\,\mathbf{j} + 2\,\mathbf{k}$

$v = \sqrt{1 + 5t^2}$

$\mathbf{T} = \dfrac{(\cos t - t\sin t)\,\mathbf{i} + (\sin t + t\cos t)\,\mathbf{j} + 2t\,\mathbf{k}}{\sqrt{1 + 5t^2}}$

11. $y(x) = x\tan\theta - x^2\left(\dfrac{g}{2v_0^2\cos^2\theta}\right)$

Chapter 9

1. 1.75 kg
2. 0.34 slugs
3. 1.12 kg

4. (a) 0.51 kg,   (b) $N = 2.9$ N, $F_{ext} = 4.1$ N,   (c) 3.7 m/s^2

5. $v(t) = -g(\sin\theta + \mu_k \cos\theta)t$

6. $t = \sqrt{\dfrac{2d}{g(\sin\theta + \mu_k \cos\theta)}}$

7. $T = 5.9$ N

8. $x(t) = 2\cos\omega t + \dfrac{1}{\omega}\sin\omega t,\quad v(t) = -2\omega\sin\omega t + \cos\omega t$

9. $E = \cos^2\omega t(2k + m/2) + \sin^2\omega t(k/2\omega^2 + 2m\omega^2)$
   $\qquad + \cos\omega t\sin\omega t(2k/\omega - 2m\omega)$

10. Critically damped

Chapter 10

1. 13.5 rad/s^2
2. 306
3. 10.7 m/s, 177 m/s^2

4. -94.2 rad/s^2, 7.1 m/s, 199 m/s^2 5. 23.4°

6. 3.8 rad/s   7. -568 N-m

8. 986 N-m-s   9. 59°

10. $\tau = 2.4\mathbf{i} - 12.0\mathbf{j}, \mathbf{L} = 40\,\mathbf{k}$

Chapter 11

1. 1.9×10^{32} J
2. 84.5 kJ
3. 1750 ft-lb

4. 30 ft-lb
5. 6.6 kJ
6. 3.3×10^6 m/s

7. 1.3×10^{-22} kgm/s, 7.7×10^5 m/s

8. 0.17 MW
9. 87 percent
10. -29.7 m/s

Chapter 12

1. 5 m, 31.4 m, 0.02 s   2. $v_p = \alpha k + \beta,\ v_g = 2\alpha k + \beta$

3. 49 m/s   4. 0.13 s   5. 2 m

Chapter 13

1. $m\ddot{x} + 2x = 1$

2. $L = \dfrac{1}{2}m(\dot{r}^2 + r^2\dot{\theta}^2) - r\theta$

3. $m\ddot{r} + \theta - mr\dot{\theta}^2 = 0$
   $mr^2\ddot{\theta} + 2mr\dot{r}\dot{\theta} + r = 0$

4. $T = \dfrac{mR^2}{2}(\dot{\theta}^2 + \Omega^2\sin^2\theta)$

5. $\dfrac{mR^2}{2}(\dot{\theta}^2 + \Omega^2\sin^2\theta) - mgR(1 - \cos\theta)$

6. $\ddot{\theta} + \sin\theta(g/R - \Omega^2\cos\theta) = 0$

7. $H = \dfrac{p^2}{2ml^2} - mgl\cos\theta$

Final Exam Answers

1. 3

2. $\sqrt{8}$

3. 8

4. $\vec{A} + \vec{B} = 2\hat{x} - 4\hat{y} + 3\hat{z}$

5. $\nabla T = \cos y\,\hat{x} - x\sin y\,\hat{y}$

6. 151 kg

7. 36 m/s

8. $\mathbf{M} = \mathbf{r} \times \mathbf{F}$

9. 187 N, 37 N, 112 N

10. 1 N-m

11. 7 N

12. Directed up

13. $\bar{a} = \dfrac{\sum M}{R}$

14. 39 N

15. 13 N

16. 50 N-m

17. 3.8 m

18. $\sum \mathbf{F} = 0$

19. $\sum \mathbf{M} = 0$

20. $W = 196$ lb

21. 2.5×10^{-7} N

22. 1600 km

23. $\phi = \dfrac{GM}{\alpha}e^{-ar}$

24. $\mathbf{g} = -\dfrac{2GM}{r^3}\left(\dfrac{3\cos r}{r} + \sin r\right)\hat{\mathbf{r}}$

25. $v = \sqrt{\dfrac{2GM}{R}}$

26. $M = mx$

27. $m_1 x_1 + m_2 x_2 = 0$

28. $X = \dfrac{\sum\limits_{i=1}^{n} m_i x_i}{\sum\limits_{i=1}^{n} m_i}$

29. 5.82 m

30. $I = \displaystyle\int r^2 dm$

31. $k = \sqrt{\dfrac{I}{m}}$

32. 4 MN

33. 670,198 lb

34. 215 ft

35. Hyperbolic cosine

36. Horizontal

37. 65 ft

38. 354 lb

39. In the opposite direction.

40. The initiation of movement.

41. $N = mg$

42. $f = \mu N$

43. $T_1 = T_2 e^{\mu \alpha}$

44. 0.04

45. 4.8 in

46. $a = \dfrac{d^2 x}{dt^2}$

47. 2.5 s

48. 37 m/s

49. 3.5 s

50. -23 ft/s^2

51. 1149 ft

52. 8.5 s

53. $\mathbf{r} = x\mathbf{i} + y\mathbf{j} + z\mathbf{k}$

54. $\mathbf{v} = -\sin t\,\mathbf{i} - \cos t\,\mathbf{j} + 2t\,\mathbf{k}$

55. $\mathbf{a} = -\cos t\,\mathbf{i} + \sin t\,\mathbf{j} + 2\,\mathbf{k}$

56. 3 m/s

57. $k = \left|\dfrac{d\mathbf{T}}{ds}\right| = \dfrac{|d\mathbf{T}/dt|}{ds/dt}$

58. $\mathbf{v} = 6t\,\mathbf{i} + 2\,\mathbf{j}$

59. $\mathbf{T} = \dfrac{1}{\sqrt{36t^2 + 4}}(6t\,\mathbf{i} + 2\,\mathbf{j})$

60. 4.9 s

61. 188 m

62. The angle of elevation should be $\theta = \theta_{\max} = 45\,°$.

63. 233 m/s

64. 36 s

65. 12.5 m/s

66. 10.5 m / s^2

67. 34 kg-m/s

68. $\vec{P} = $ a constant (for an isolated system)

69. $\dfrac{d\vec{P}}{dt} = 0, \Rightarrow \sum F_{\text{ext}} = 0$

70. $\dfrac{d\vec{v}}{dt} = 0$

71. $1\,\text{N} = 1\,\dfrac{\text{kg-m}}{\text{s}^2}$

72. $F_{\text{kin}} = \mu_k N$

73. $\vec{F}(\vec{r}) = -\vec{\nabla} U = 2re^{-r}(r - 2)\hat{r}$

74. $v(t) = -\sqrt{\dfrac{2\alpha}{m}}\sin\left(\sqrt{\dfrac{2\alpha}{m}}t\right)$

75. 3.5 N/m

76. $\omega_0^2 = \dfrac{k}{m}$

77. $T = 2m\,\omega_0^2 \sin^2 \omega_0 t$

78. 4×10^{-5} joules

79. $\dfrac{d^2 x}{dt^2} + \omega_1^2 \dfrac{dx}{dt} + \omega_0^2 = 0$

80. A decaying exponential.

81. The underdamped case oscillates, but the oscillation decays in time.

82. Toward the center.

83. $15.7\,\dfrac{\text{rad}}{\text{s}^2}$

84. 373

85. $T = \dfrac{1}{2}I\omega^2$

86. 4891 N-m

87. $\mathbf{M} = 9t^2\mathbf{i} + 7\mathbf{j} + \sin t\,\mathbf{k}$

88. $\tau = 8.0\mathbf{i} - 8.0\mathbf{j}$

89. $1\text{ eV} = 1.6 \times 10^{-19}\text{ J}$

90. 753,519 ft-lb

91. 366 N-m

92. Positive

93. 67 percent

94. $8k$

95. $9k$

96. $\ddot{x} + \dfrac{15}{2m}x^4 = 0$

97. $L = \dfrac{1}{2}m\dot{x}^2 - \beta x$

98. $\ddot{x} = -\dfrac{\beta}{m}$

99. $p = m\dot{x} + \alpha$

100. $H = \dfrac{1}{2}m\dot{x}^2 + \beta x^2$

Bibliography

F. P. Beer et al., *Vector Mechanics for Engineers, Statics and Dynamics*, 7th ed., McGraw Hill, New York, 2003.

D. Halliday and R. Resnick, *Fundamentals of Physics*, 3rd ed., Wiley, Massachusetts, 1988.

E. W. Nelson, C. L. Best, and W. G. McLean, *Schaum's Outlines Engineering Mechanics: Statics and Dynamics*, 5th ed., McGraw Hill, New York, 1998.

REA's Physics Problem Solver, Research and Education Association, Pascataway, NJ, 1990.

S. K. Stein, *Calculus and Analytic Geometry*, 4th ed., McGraw Hill, New York, 1987.

M. Thornton, *Classical Dynamics of Particles and Systems*, 4th ed., Saunders College Publishing, Philadelphia, 1995.

INDEX

Statics and Dynamics Demystified

240